The Essential Guide

VINO

to Real Italian Wine

Joe Campanale

WITH JOSHUA DAVID STEIN

PHOTOGRAPHS BY
ODDUR THORISSON

CLARKSON POTTER/PUBLISHERS
NEW YORK

For my partner, Ilyssa;
our rescue pit bull, Susu;
and our son, Cole

CONTENTS

INTRODUCTION

"There's never been a better time to drink Italian wine." By this point, this line has been written so often that it has become a cliché. It may seem strange to start a new book with a cliché, but what was true when it was written by Burton Anderson in the 1980s, or by Joe Bastianich and David Lynch in their pivotal 2000 book *Vino Italiano*, is still true today. That is the signal trait of evolution: constant, steady improvement.

I wrote this book to capture this particular moment in the evolution of Italian wine, and to celebrate the producers who have been responsible for pushing it forward. As you'll see in the following pages, the coming together of a few interrelated but distinct movements has resulted in a flourishing of truly incredible Italian wines. These wines, in my opinion, deserve a new way of thinking about, or framing, Italian wine, an approach that takes what is in the bottle as just one chapter in a longer, more profound story. Thinking about wines in this way, in their cultural, geological, and human context, has profoundly deepened my enjoyment and appreciation of them.

Over my close to twenty years in the restaurant business, I've lived, breathed, and drunk Italian wine. I've built scores of Italian-focused wine lists, tasted thousands of bottles, and traveled to Italy dozens of times to meet with the country's most fascinating and influential producers. I've been a forager, hunter, explorer, advocate, importer, cheerleader, and storyteller of Italian wine. Mostly, though, I've been a student of it. And from this vantage point, I've seen how Italian winemakers have fully come into their own over the past two decades. Producers first, and then later, in most cases, the markets have turned away from the pat wisdom of the international varieties, the ease of high-volume industrial viticulture, and the reliance on the traditional Italian wine pyramid as the ultimate indicator of quality. They've come to see the value in their own native grape varieties and have reoriented themselves to focus on how to best express the terroir of their own country, their own hometowns and hills, rather than mimicking so-called international styles of wine. I've seen how they have been part of a more global movement toward sustainability and minimal intervention, and I've tasted how the fruits of this labor more than justify it.

It's time to reexamine what Italian wine is now and, importantly, what it could be, for both have greatly expanded. That's the idea behind this book: to capture a true portrait of the multihued and dynamic world of contemporary Italian wine. In the first part, I lay out the essential characteristics of what I've looked for over the last fifteen years or so as wine director for many Italian-focused restaurants. First I deconstruct the outdated wine pyramid of DOC and DOCG, the governmental designations that have defined Italian wine for the last half century, and then explain my own rubric of quality, the Vino Vero Venn diagram. The VVV, as I call it, plots how a great wine should embody three overlapping values: native grapes, exceptional terroir, and an artisan winemaker well versed in low-intervention methods that can allow those grapes to shine. Although not every wine occupies the exact center of the Venn diagram, the center is the ideal. The VVV isn't, perhaps, as geometrically imposing as the wine pyramid, but I have found that it is much more helpful as a tool for discovering and thinking about vino vero.

In the second part, I explore a series of exciting new trends, from pet-nats to orange wines to aged white wines and more, that have resulted from the renaissance of artisan wine in Italy.

And in the third part, the lion's share of the book, I take you on *un grande giro*, embarking on an epic quest of Italy's twenty regions, uncovering and championing the wonderful wines and producers in each. (There are wonderful wines and producers in every region, no matter how small.) In each section, we begin with the context—historical, cultural, geographical, political—that has given each region its unique character, one that is as rich and textured, as complicated and refreshing as its own inhabitants. The differences among the regions range from subtle to outstanding and have as much to do with pedology (the study of soil) as they do with the ebb and flow of ancient history. I describe the charms and characteristics of, for example, the steep hills of the Valle d'Aosta, or the near-tropical climes of Sicily, connecting how the relationship between the terrain and those who navigate it has come to shape the wines that make it into the bottle. In the second section, I showcase the grapes native to each region. (There is, of course, substantial overlap among regions.) Although the names of some of these varieties are rightfully well known, many are still almost criminally obscure. For me, some of these latter types are among the most interesting. And, finally, I visit some of the wonderful producers I've met over the last twenty years—many in person, some only through their wines. These men and women (and, happily, today there are more and more female producers receiving the recognition they deserve) are the beating heart of this book. It is their dedication and vision that bring forth from their hills, cliffs, valleys, and volcanic slopes vini veri. And getting to know them, at least through these pages, getting to know their stories, deepens the experience and the enjoyment of drinking their wines.

JOEY
from QUEENS

Encountering someone working in Italian wine with a name like Joe Campanale and a Queens accent that occasionally peeks out (as when I say "tawk" or "New Yawk"), people make a lot of assumptions about how I grew up. "What I wouldn't give to have been at a Sunday dinner at the Campanale household!" they say, no doubt with visions of baked ziti fairies dancing in their heads. So they're surprised when I tell them I was raised in a Jewish household by a single mother in Forest Hills, a world away from the Little Italy of Arthur Avenue or Bensonhurst. They're surprised to learn mine wasn't a childhood of red sauce and Chianti, but instead one of pastrami sandwiches at Ben's Delicatessen and knishes at Knish Nosh on Queens Boulevard. My mom, who worked at an endless stream of low-paying jobs, from shoe saleswoman to telephone psychic, to provide for the two of us, didn't have the time or money for elaborate feasts, or really feasts of any kind. (And as far as the Campanale side of the family goes: After my father died in 1986, when I was two years old, they drifted to Florida, never to be seen again.) No, my obsession with Italian wine came much later in life, and almost accidentally.

In 2002, I enrolled at New York University, carried there on a mixture of grants, a scholarship, and student debt. My plan was to become a lawyer, at least for long enough to achieve some financial stability, and then, like so many freshmen imagine, I'd go out and save the world. But during my first year, I took a class simply called Beverages, which covered wine, beer, and spirits, through the Food Studies program at NYU's School of Education. (What college freshman wouldn't take a class called Beverages?) The class met at ten a.m. and, because we were all under twenty-one and this was, after all, school, we were required to spit as we tasted. It wasn't the first time I had tasted wine, but it was the first time I had thought about it. As our professor, Linda Lawry, explained to us, each bottle contained the story of the place from which it had come. As many bottles as there were, there were stories to tell and to hear.

I was hooked. By the end of that year, I was a regular at Union Square Wines' Sunday afternoon tastings and proud of my growing body of knowledge. As my friends downed sake bombs at East Village bars that didn't card, I nursed bottles of affordable Spanish wines like Juan Gil Jumilla and Alvaro Palacios "Pétalos" in my dorm room, thrilled that I had snuck them past campus security.

The requisite next step in the career of any budding oenophile—or at least one like me, rapidly reconnecting with my Italian heritage through wine—was a pilgrimage to Italy. Luckily for me, in 1994, Sir Harold Acton bequeathed a Renaissance villa in the hills of Florence to New York University, which then established NYU Florence, a study-abroad program. So I headed to Italy, assigned to live in an apartment in the Oltrarno, a working-class neighborhood on the southern banks of the Arno River. It was the best thing that could have happened to me.

When I was a kid, my mother and I would walk past the gourmet grocery stores on Austin Street in Queens, peering in at the fancy jams with their French labels and the dark green bottles of olive oil. We couldn't afford any of it, but looking was free. In Florence, as I wended my way through the city to campus every day, I continued to look. I wandered into the macellerie, the pastificie, the pasticcerie, the alimentari, and the salumerie, marveling at how much more poetic butcher shops, bakeries, and bodegas seemed in Italy. I was entranced by the infinite variety of cornetti, as the Italians call croissants, the deli case full of *pasta fresca*, and the ceilings hung heavy with salumi.

One day on the way back to the Oltrarno, a friend and I ducked into a sleepy wineshop on Via Roma. It was one of those places that couldn't hold more than five or six customers. Behind the counter sat a distinguished-looking gentleman with salt-and-pepper hair, a soft sweater, and a nice smile. He greeted us with a *Salve!* That, in itself, was unusual. Under the assault of so many tourists and students, many Florentines curl up like pill bugs—understandably so. But this man was ready to talk, and we were ready to listen.

Pulling bottles from the shelves, he spoke to us about his favorite wines as if they were friends. (In fact, I think, many of the producers *were* his friends.) "This is Fontodi by Giovanni Manetti," he'd say. "Giuseppe Rinaldi, good to see you again." I'm not sure whether I ever knew this man's name. We called him Signore to his face and Il Professore to each other. He wasn't teaching us so much about the details of the individual bottles as modeling an approach, that each wine should be seen as an individual expression, full of personality, history, stories, and memory. The bottles he stocked were far from the industrially produced wines I usually encountered. His wineshop became a regular stop on my walks to and from campus. Meanwhile, in the evenings, my friends and I frequented *trattorie* and *osterie* like Osteria Santo Spirito and Trattoria Mario. Whatever wine I drank, I loved; when I could splurge, I did. We'd drink big, fruity Tuscans and look at each other, grasping the slender stems of our glasses, and announce, "This is good." But our praise was always half a question, as if we were trying to convince ourselves that it wasn't crazy to spend all our euros on bottles whose labels we couldn't decipher.

Still, I was thirsty for more. On the weekends, I would often rent an Alfa Romeo 147, the cheapest automatic car the rental place offered, and head into the country to visit the wineries whose vineyards turned the famous Tuscan hillsides into a patchwork quilt of vines. These wineries seemed a world away from those arcane bottles and expensive glasses of our prodigal feasts. Here farmers had dirt beneath their fingernails, and the air, heavy with the scent of grapes, sang with the chirps of birds, buzz of bees, and occasional bark of the omnipresent winery hound. A few feet away from the vines, in massive buildings, were the vats and barrels that turned the bunches of Sangiovese grapes into, mostly, Chianti. The scents of ferment and wood were as enchanting as they were unique. Back in the classrooms of La Pietra, I was learning about Italian wine from no less august an expert than Ian D'Agata, but I had never truly understood the thread that connected a land, its people, and its produce until I found myself standing in those vineyards.

One day, after a particularly rousing night at Trattoria Cibrèo, a friend and I stopped by to visit Il Professore. I told him about the bottle of a famous producer of Chianti Classico on which I'd spent pretty much my last cent. I anxiously awaited his words of praise. Instead, he shook his head sadly, put his hand over his face, and said, "*Questo, questo non è il vero vino*" (This, this is not *real* wine).

Then he walked over to one of his shelves and pulled out a bottle of Podere Le Boncie's Chianti Classico "Le Trame." He ceremoniously opened it and solemnly set out three glasses. "*Allora, eccoci, un vino vero*" (Here, here is a true wine).

He poured some of the scarlet wine into the glasses, and I immediately perked up. It was like no wine I had smelled before. My nostrils filled with its earthy, ever-changing scent, so unlike the big-nosed fruit-forward Chianti I was used to. As we drank it, Il Professore told us the story of Le Trame's winemaker, Giovanna Morganti. He spoke of how the wine was fermented in open vats and how Giovanna grew the grapes on her farm in a small town called San Felice using biodynamic methods and minimal intervention. But mostly, he just let us experience everything this wine—a vino vero—could be.

From that day on, whenever we went out to eat—which was always—I'd ask for *una bottiglia di vino artigianale*. And then I'd usually have to clarify, "*Una bottiglia di vino artigianale meno costoso*"—a less expensive bottle of artisanal wine.

I returned to New York that summer with my hair on fire about wine. While my friends were getting internships at law firms and magazines, I managed to finagle one at Italian Wine Merchants, on Union Square. IWM, as it was called, was one of the first wine stores in New York City to champion Italian wine. Though the scene in New York was changing, thanks to restaurants like Babbo and il Buco, French wines were still predominant. The dine-out set did drink wines from Italy, of course, but these were never given the respect of those hallowed French names like Bordeaux and Burgundy. IWM, on the other hand, was already carrying the orange wines of Gravner and Damijan, and bottles of Bartolo Mascarello's Barolo, names that are popular today but were nearly unheard of then.

Still under the legal drinking age, I worked the floor among a cadre of Italian wine aficionados, helping to restock the shelves and work the register. After hours, I'd assist in curating the tasting events held in the back room. The staff got a kick out of this young kid obsessed with Italian wines, and they'd send me down rabbit holes researching esoteric grapes like Ribolla Gialla and Sagrantino.

After a year, I was offered a job at the International Wine Center, an esteemed educational institution where Linda Lawry, my old professor, was now director. I was sad to leave IWM, but the added bonus was that my new employers would pay for my diploma courses at the Wine & Spirit Education Trust—which cost upward of $4,000. Meanwhile, I was still going to NYU, though I had discarded the dream of working at a white-shoe law firm. Instead, I was racking up graduate credits in Food Studies, while working multiple on-campus jobs in addition to my wine gig to help pay tuition.

In 2006, I graduated from NYU, and two years later, I earned a master's degree in Food Studies there, along with a diploma from the Wine & Spirit Education Trust. Now I was out in the real world. My old boss at Italian Wine Merchants, August Cardona, called me and said Babbo was looking for a sommelier. I had never worked as a somm, and I wouldn't even call myself one, though I had passed the Certified Sommelier course from the Court of Master Sommeliers. (To me, a certificate is a necessary but not sufficient prerequisite.) But Babbo was then the mecca of Italian wine, and the chance to work with David Lynch, the legendary wine guy, was too good to pass up. During my interview with him, I quoted passages to David from his own seminal book, written with Joe Bastianich, *Vino Italiano*, which I had largely committed to memory, and I walked away with the job.

As a junior somm, the lowest on the totem pole, I had the more difficult second floor of the restaurant, approximately sixteen tables, as my territory. The key to success was cardiovascular endurance. Babbo occupied an old Greenwich Village townhouse and the wine cellar could only be reached by running down the flight of stairs to the first floor, squeezing between the bar and the host stand and descending another steep flight into the basement, and then wending one's way to the very back of the building. The cellar itself, though full of exceptional wine, was a poorly organized mess. After rifling through boxes and crates, I'd rush back up the stairs to the table. My suit—proudly purchased at Banana Republic with the last of my savings—was invariably soaked with sweat by the end of the night.

This was, by the way, 2007, at the feverish apex of New York's spendy aughts. Babbo was filled with high rollers who didn't bat an eye dropping $250 for a bottle of Giacomo Conterno Barolo or Poggio di Sotto Brunello. I was serving the 1 percent, but I was certainly not among them. After my shift, when I'd go out to drink wine with my industry friends, the bill would be a moment of reckoning, a rude awakening at one a.m., as I signed away what seemed like a large chunk of my East Village rent. The whiplash from pleasure to pain smarted, and my conviction grew that there was no reason why the wonderful Italian wines I had enjoyed in Italy and in New York—wines created by artisans, by farmers, essentially by artists—should not be affordable to artists and artisans (and students and misfits and the marginal and the middle class) too.

When August decided to venture out on his own from Italian Wine Merchants to open a restaurant a few months after I'd arrived at Babbo, he asked me if I wanted to join him as a partner and wine director. I was only twenty-three years old, but I jumped at the chance with the confidence one has only when he (or she) is twenty-three years old. When we opened dell'anima, which means "of the soul" in Italian, a tiny forty-two-seater in the West Village, I had the opportunity to craft a wine list that embodied all I had been thinking about: an appreciation for artisanal wines, passed on from Il Professore; a love for native grapes, the result of my classes with Ian D'Agata; an interest in organic methodology, picked up while earning my master's in Food Studies; and, of course, my exposure to some of the highest-quality Italian wines at IWM and Babbo.

The list was 125 wines strong, lengthy for so little a restaurant, with a heavy emphasis on approachable wines by the glass. Instead of slouching toward what I saw as the clichés of Italian wine, I wanted to steer guests toward new discoveries: I was adamant that there would be no Pinot Grigio or Prosecco, mass-produced wines that at the time seemed to be cynically designed to be drunk rather than appreciated. So, instead of a Pinot Grigio, I'd recommend Ermes Pavese's Blanc de Morgex, a wine from a two-hectare vineyard on the Italian side slopes of Mont Blanc; in lieu of a glass of Prosecco, I'd offer a Lambrusco Bianco from Lini 910, a fourth-generation winery currently run by a trio of siblings. Though I hadn't come up with the Vino Vero Venn at the time, both of these wines would have occupied the prime center plot: They were made with native grapes on exceptional terroir by skilled artisanal producers. Intimate and affordable, dell'anima was an instant hit. The next year, when we opened L'Artusi, I created a wine list organized by the twenty regions of Italy, which allowed our guests to travel beyond Tuscany, Piedmont, and the Veneto to lesser-known provinces like Molise and Calabria. Two years after that, we opened Anfora, a straight-up wine bar, where I dove deep into producers, offering not just one or two but five or six wines from winemakers I really loved so that our guests could better understand a particular producer's philosophy as expressed through his or her wine. And in 2012, we opened L'Apicio, a vast restaurant in the East Village that synthesized the focused approaches of L'Artusi and Anfora, but in a much bigger, splashier space.

The businesses were booming, but in 2016, it was time for me to move on. A few years earlier, I had moved to Brooklyn with my then-partner, and now also the mother of my child, and fallen in love with the borough. After I'd spent months looking for a place to open a restaurant, my friends Francine Stephens and Andrew Feinberg, who ran the lauded pizzeria Franny's on Flatbush Avenue, asked whether I'd be interested in taking over their lease. Franny and Andrew had had enough of the New York City restaurant grind. They had two kids and wanted more space and fewer worries, so they were planning on closing Franny's and moving to a farm in Vermont.

Franny's was my favorite restaurant, a perpetually packed pizzeria with tasty market-driven contorni and a compact, well-curated wine list by a Florentine named Luca Pasquinelli, formerly the senior sommelier at Babbo and Il Latini. That winter, my partner, Erin Shambura, the former executive chef at L'Artusi, and I opened the doors of Fausto, a restaurant all our own.

Erin turned out fresh pastas, her own expertly made contorni, and wood-fired meats while I poured from a wine list made up exclusively of wines I wanted to drink at home. They were, like Il Professore's wines, as close to friends as a bottle could be. They were drawn from Italy, primarily, but also included some from select producers in France, Spain, and the United States. I knew most, though not all, of the winemakers personally. I had walked through their vineyards, eaten at their farmhouse tables, and listened eagerly to their stories. So when diners came into Fausto, I loved to decant not just a surprising bottle of orange wine, say, but also the story behind it. I liked to think, long after the bottle had been emptied, that is what they took home with them.

The following year, in a small storefront a few blocks away on Vanderbilt Avenue in Prospect Heights, which was fast becoming Brooklyn's Restaurant Row, we opened LaLou, a wine bar that allowed me to further explore the world of natural wines. Luckily, the natural wine world had greatly expanded since my first bottle of Le Trame in 2004. In the last twenty years, many more wineries around the world—including, of course, Italy—have tilted toward native grapes, minimal intervention, and individual expression.

PART I

What Makes a Vino Vero?

The
Italian Wine
Pyramid

For nearly the last sixty years, a mere drop in the bucket when it comes to the history of Italian winemaking but nearly the entirety of its presence in the American consciousness, Italian wine has been organized by a system of appellation called Denominazione di Origine Controllata (DOC), or, more colloquially, the wine pyramid. This system, inspired by France's Appellation d'Origine Contrôlée (AOC), sorts wines into four increasingly regulated categories, based largely on geographic origin, although other rules cover everything from grape variety to yield per hectare (the volume of wine produced per 1/100th acre) to methods of vinification (how a wine is made), minimum alcohol levels, and duration of aging.

When I first began learning about Italian wine, I regarded the pyramid as sacrosanct. As one moved from the lowly VdT to the rarefied quarters of DOCG wines, I thought, naturally, that the quality of the wine improved as well. I took

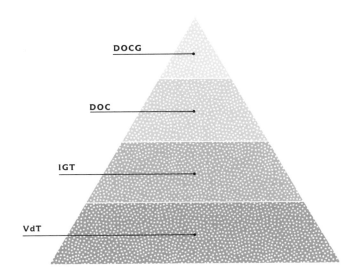

DOCG

DOC

IGT

VdT

Italian Wine Classifications

If you've picked up this book, you may already be familiar with at least the general contours of the Italian wine pyramid, but I'll quickly outline them here.

VINO DA TAVOLA (VdT)

In "Scenes from an Italian Restaurant," when Billy Joel sings, "A bottle of white? A bottle of red? Perhaps a bottle of rosé instead?" he's probably thinking of *vino da tavola*, or table wine. Introduced in 1963, VdT is the lowest tier and the least regulated category in the pyramid. It's basically a free-for-all and, traditionally, it has been priced at nearly free for all. The grapes can be sourced from anywhere in Italy, with no strictures regarding hectare yield, vinification, or additives such as sugar (in a process called chaptalization) or wood chips (often added to simulate oakiness). Neither vintage nor variety can be included on the label. In fact, the labels of VdT wine are by law almost comically basic: Vino Bianco, Vino Rosato, or Vino Rosso.

When the pyramid was first introduced, most VdT wine didn't even come in bottles but was instead sold in bulk. Historically, it rarely made financial sense for these wines to be exported, so the majority of them were consumed within Italy and the EU. Today, VdT wines account for approximately 35 percent of the 47,500 hectoliters produced annually.

INDICAZIONE GEOGRAFICA TIPICA (IGT)

Moving up the pyramid, you arrive at the slightly more proscribed category of IGT. A relatively recent renovation, it was added to the pyramid in 1992 to act as a bridge between the Wild West of the VdT and the straitlaced domains of DOC and DOCG. While much about IGT wines is still free-form in terms of methodology, as the name suggests, an IGT must contain grapes from a specific geographic region and that region must be indicated on the label. In addition, there are limits to yield per hectare (how densely grown the land is), but these are so high that they aren't really a factor. Today, there are 118 IGT wines, accounting for approximately 24 percent of the total Italian production.

DENOMINAZIONE DI
ORIGINE CONTROLLATA (DOC)

Scaling the pyramid, the rules get tighter and factors like allowable maximum yield, minimum alcohol level, and length of aging more circumscribed, with the noble goal of producing a wine that tastes like where it's from. This starts in earnest at the DOC level, which is also the most well known of the designations. (It should be: It's the title track.) Here, grape varieties, minimum alcohol levels, acidity and extract levels, maximum yields, winemaking techniques, and aging are all minutely stipulated. There are currently 333 DOC winemaking regions, and the number seems to grow yearly. Regulations naturally vary in their intensity across these regions due not only to the myriad varieties of wine but also to cultural and historical reasons, the strength and maturation of the winemaking tradition in the area, and, of course, politics. In some areas, methods of acidification and irrigation are also stipulated. Unlike wines lower on the pyramid, all DOCs must pass an organoleptic, or sensory, exam by a government panel before receiving certification.

DENOMINAZIONE DI ORIGINE
CONTROLLATA E GARANTITA (DOCG)

If DOC is the Gold Card of Italian wine, DOCG is the Platinum. At the apex of the pyramid, DOCG wine "enjoys" all the restrictions of its slightly more permissive cousin as well as various additional regulations. It is these add-ons that yield the G (or guarantee) at the end of the category's name. Chief among these more difficult hoops are *two* organoleptic exams, tasting panels that the wines must pass before being awarded the DOCG seal from the Ministry of Agriculture.

Although the DOCG designation was created at the same time as the DOC, the standards for meeting it were so rigorous that it wasn't until 1980 that the first wines—three from Tuscany, including Brunello di Montalcino, Vino Nobile di Montepulciano, and Chianti, and two from Piedmont, Barolo and Barbaresco—were awarded the status. Today there are seventy-seven DOCG appellations, the majority of them still centered in Piedmont and Tuscany.

that as dogma, and I wasn't alone. (Human beings crave structure, a scaffolding on which to hang the messiness of the world, not just in wine but in all things.) The wine pyramid was a clean, elegant, and presumably accurate way of thinking about the complicated world of Italian wines. I, and many others, took it to be immutable, a monument—and, like a monument, not to be questioned.

But, as I came to find out, I was mistaken. Had I excavated the area surrounding the structure, I would have seen that, like all monuments, the wine pyramid was built during a specific time, and in a specific place, to safeguard or serve the interests of a specific constituency. In 1963, Italy was a country in motion. World War II—and World War I before it—had laid waste to the countryside and its economy. The nation itself, an uneasy alliance of duchies, kingdoms, principalities, and republics, was barely a century old. In the early twentieth century, despite the country's relatively small size, traditions, allegiances, terrain, and even languages varied greatly throughout Italy. Famine and unrest in the rural south, poorer than the north to begin with, had caused great waves of postwar migration. The southerners made their way to northern cities like Milan—which, in turn, fueled already-inflamed regional resentment—and to the United States, emptying the countryside of a much-needed workforce (and sending my father's family to Ellis Island).

However, thanks largely to America's Marshall Plan (the United States saw left-leaning Italy as a key battleground in the Cold War), which injected billions of dollars into the economy, Italy began to enjoy twenty years of sustained economic growth and rapid industrial expansion, growth it hasn't seen since. This *boom economico* was consolidated with the signing of the Treaty of Rome in 1957, which established the European Economic Community, the forebear of the European Union, and exponentially expanded trade among the six signatory nations (Italy, West Germany, France, the Netherlands, Belgium, and Luxembourg). The result of all of this was that by 1963, when Law No. 930, establishing the framework of the DOC, was passed by the Italian Senate, the country and its lawmaking apparatus were in the service of industry, including industrial winemakers.

At the time, the vast majority of the wine produced was low-cost bulk wine destined for Italian tables. But with a common European market established, Italian producers were concerned with counterfeiters and competition. Taking a page from France's AOC system, the winemakers wanted to protect themselves by enshrining into law the status quo of how they made their particular wines in that moment. This, they believed, was A Very Good Thing, since it would protect their brands from imitation and, just as the AOC had in France, bestow upon their wines in the international market the respect afforded an official designation. And they were right, but the laws also served to encapsulate in amber the practices of those mid-century industrial winemakers who held the lion's share of power in 1963. The language of the DOC regulations is full of phrases like "in conformity with existing practices" and "so as not to change the nature of wine." This allowed industrial wine producers like Bolla, which churned out millions of gallons of Soave (accompanied by the catchy jingle, "There's more, more, so much more than wine in a bottle of Bolla"), to gain valuable market share and consolidate their power. Bolla was right: There's also profit to be considered, and intra-European competition, and corporate influence upon trade legislation. But for many bottles of wine, quality was not a question to be considered.

Almost immediately upon the introduction of the DOC system, its structure began to collapse. Enterprising winemakers who, like many Italians, prided themselves on their *furbizia*, or cunning, immediately found ways to work around and undermine the system in order to make wines that were to their own taste. Wines that were too cloudy, or too light in color, or, bizarrely, had too high a percentage of single grape variety failed to gain DOC status. That these winemakers produced wines of such stellar character that nevertheless failed to meet DOC standards made clear how flawed the system was. And for many others, the obeisance to industrially produced wine ran counter to everything they understood wine meant. As the influential anarchist wine critic Luigi Veronelli put it, "*Il peggior vino contadino è migliore del miglior vino industriale, perché ha un'anima*" (The worst wine of a farmer is better than the best industrial wine, because it has a soul).

What followed were years of alteration and jerry-rigged renovation, along with attempts by well-meaning authorities to make a flawed system better. Political pressures exerted their destructive force on the pyramid too. Local and regional governments that wanted a larger slice of the Italian wine market, and could do so by collecting higher taxes on more expensive wines, lobbied for their own DOC and DOCG designations. Some of these, like Fiano di Avellino and Cerasuolo di Vittoria, made sense. Others, like the 1987 elevation of notably mediocre Albana di Romagna to DOCG, were examples of straight-up viticultural gerrymandering. The ongoing absence in that category of others like Etna Rosso and Etna Bianco, which are both mere DOCs, makes a mockery of the designation. In 1992, Giovanni Goria, the minister of agriculture, pushed through the so-called Goria law, which further subdivided regions and subregions and clarified the framework considerably.

But by then, some truly great winemakers had already decided to opt out of the race to DOC or DOCG status, and that trend of either dropping out or not even applying for inclusion has only accelerated in the intervening years. Thus, some of the best wines have held on to the "lowly" VdT and IGT designations so their makers can be as creative and innovative as the wines deserve. I always think of Peter Weimer and Romy Gygax at Cascina Ebreo, whose unfiltered 100 percent Nebbiolo Barolo failed to pass the organoleptic exam to qualify for a DOCG due to its cloudiness, or turbidity. (Filtering, and thus clarity, being mistakenly associated with quality.) So Peter and Romy named their wine "Torbido!" and shunned the appellation, continuing to make their terrific idiosyncratic wine, with *un'anima*, as Luigi Veronelli would say.

Today the pyramid is in ruins. Though many of my favorite wines still wear the badge of DOC or DOCG certification, its utility as a measure of quality has evaporated, if it ever existed in the first place. That, I think, is a good thing. Too often we cling to a structure because it's a structure, without examining the integrity of its foundation, or considering who it protects and who it excludes. Though the DOC system will continue to be helpful as a broad organizational hierarchy, and has certainly been invaluable in proving the worth of Italian wines internationally and even protecting varieties that might otherwise have disappeared, like all works of man, its time of obsolescence has come.

The
Vino Vero
Venn

But if not a pyramid, then what? Structures are useful. So instead, I propose an alternative: the Vino Vero Venn diagram. It may not have the geometrical zip of a pyramid, but focusing on the intersection of three essential factors—native grapes, exceptional terroir, and artisan winemakers—has guided my approach to Italian wine (and non-Italian wine, for that matter) for years and has not failed me yet. Before we get started, though, I want to make explicit the end goal of the VVV (also called—by no one for now, but maybe that will change—the Triple V).

Unlike the pyramid, the VVV isn't meant to freeze the wine world at one moment in time, to enshrine it as a Platonic ideal, and to use it as a measure for all future wines. Instead, it's meant to guide us to wines that faithfully reflect and embody their *terroir*, or land. (Much) more on this later. Underlying this approach,

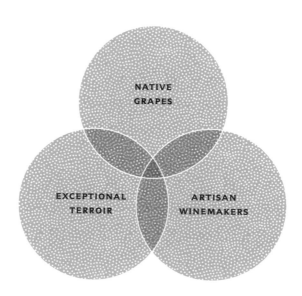

of course, is the assumption that each individual factor—the land, the vines, and the winemakers—is of high quality and worthy of veneration. As Luigi Veronelli, wrote, "*Il vino, dopo l'uomo, è il personaggio più capace di racconti*" (Wine, after man, is the character most capable of stories). If Veronelli is right, then the purpose of winemaking is to give voice to as many stories as possible, told with as much skill as possible, capturing a world as richly as possible, with wines as complex, varied, and satisfying as the land and people from which they come. That is the goal, and, I think, a noble one.

The VVV would be of only abstract interest and limited use if the criteria were so difficult that few wines qualified. By intent, that is not the case. Over the last twenty years, there's been a powerful sea change toward embracing smaller vineyards with native grapes. Its heroes are winemakers who have embraced the challenge of a light touch, a personal vision, and a deep connection to their land. As the three circles of the VVV converge, that bull's-eye is growing larger and larger. Among the greatest joys in my life have been the discoveries I've made, friends I've met, and wines I've tasted following its criteria. I hope the same is true for you.

NATIVE GRAPES

I grew up in and still live in New York City, where it's a common pastime to bemoan how homogenized everything's become and to yearn for the "good old days." It is true that the businesses that now line the blocks of much of Manhattan can seem a bit cut-and-paste: a bank, a CVS, a Starbucks, and, maybe, if you're lucky, a bodega or a restaurant. Most of the gourmet groceries on Austin Street, down which I used to stroll with my mother, have been replaced by big-box stores with global footprints. It's a bummer. What I've always loved about New York—and what everyone loves about their own hometown—is what could *only* grow here: the appetizings, the delicatessens, the oddball repair shops, the fountain pen hospitals, and, more recently, high-end anime stores, places that seem to draw their lifeblood from the concrete below them.

Why would wine be any different? A grape is a small orb that holds within its skin millennia of history. A vino vero is one that uses grapes native to its land, grapes that perhaps could not (and, at any rate, didn't) develop anywhere else. (And, yes, this leaves New World wines at a disadvantage, since we have so few native grapes, but nothing is black and white. Such is the nuance of the VVV that two out of three ain't bad. On the other hand, it is why I love Italy so much.) According to Ian D'Agata's encyclopedic *Native Wine Grapes of Italy*, the country grows approximately two thousand endemic varieties. That means Italy has more native grapes than France, Spain, and Greece combined.

And yet such was Italian winemakers' lack of confidence and thirst to enter the international market that, beginning in the 1970s, with the runaway success of the Super Tuscans—wines made in Tuscany with French grapes—winemakers across the country planted rows and rows of Merlot, Cabernet, Sauvignon Blanc, and Syrah to soften the edges of their wines with what they saw as more market-friendly French grapes. (It should also be noted, of course, that in some parts of Italy, notably the northern regions of Friuli Venezia Giulia, Trentino–Alto Adige, and the Veneto, French grapes have been grown since the eighteenth century. But then again, in the eighteenth century, Italy didn't yet exist as a country.) This globalization of grapes yielded cookie-cutter wines, pleasing perhaps on the palate but lacking a clear expression of the land.

More recently, though, in the last twenty years or so, scores of winemakers have been returning to their native grapes—and not always the most well-known ones. Some, like Timorasso, the rare Piedmontese white that grows in the rolling hills of the Colli Tortonesi, are being made into stand-alone wines for the first time, thanks to growers like Walter Massa, who almost single-handedly rescued the variety from extinction. In fact, in Piedmont, where Nebbiolo was once the only star, other grapes, like ensemble cast members—including Pelaverga, Freisa, Ruché, Grignolino, Arneis, and Favorita—are stepping into the spotlight.

All across Italy's twenty regions, producers ranging from Elisabetta Foradori of Trentino, savior of Teroldego, to Heinrich Mayr of Nusserhof, in nearby Alto Adige, hero of Blatterle, to Joy Kull of La Villana in Lazio, who's working with a local nursery to plant over a dozen heirloom grapes in order to determine what will do best on her land, are championing these native grapes. They're all animated by an almost religious zeal, a bedrock faith that wine should tell a story of a place, a plot, a history—and that the language with which to do that is a native grape.

EXCEPTIONAL TERROIR

Like *saudade* in Portuguese, *chutzpah* in Yiddish, or *han* in Korean, there's no exact translation for the word *terroir*. It comes from the French word *terre*, meaning land, but terroir signifies not just the land but the sum total of the land. And when it comes to wine, terroir also picks up an almost mystic undertone as the spirit of the land. That this spirit can be expressed in a bottle is what makes wine magical. Without terroir, wine is just a beverage.

Francesco de Franco of À Vita

Not all land is equal, of course, especially as it pertains to grapes, and that makes terroir infinitely fascinating. The terroir of flatlands is different from the terroir of a slope. Up on the slope, you must ask: What is the angle at which the breeze flows? What is the provenance of that breeze—is it a sirocco or a ponente or a bora? How and when does the sun strike the plants and for how long? Is the soil loamy or gravelly? Is the dirt black or brown? These are all parts of what determines the character of a plot of earth, the *terroir*.

By the time a bottle reaches your table, it's hard to imagine that the liquid inside was so profoundly influenced by, say, whether the vines were 100 or 500 meters above sea level, whether they were exposed to light from nine a.m. to five p.m. or from eleven a.m. to seven p.m., whether a certain terrace wall protected them from a Mediterranean breeze or that wall had fallen and they were exposed. But such conditions are the memories of the wine. And just like the memories of your first kiss or the first time your heart was broken, they have an existential impact.

Thanks to ancient geographical and tectonic occurrences, Italy has a terrific and truly astonishingly varied terroir. From the sub-Alpine slopes of Valle

d'Aosta to the dry, hot volcanic island of Sicily, which shares the same latitude as North Africa, there are endless combinations of light, soil, wind, heat, shadow, precipitation, and elevation. To quote Mel Brooks on talent, "You either got it or you ain't." And Italy's got it.

Terre, earth, is the raw material. It's pure potential. To become terroir, at least in wine, it needs a winemaker with the skill and vision to harness it. The strength of many intergenerational winemakers, like Maria Teresa Mascarello and Francesco Valentini, isn't their heraldic coat of arms or impressive granite *castelli* but the fact that they have worked the same land for years and years and know its characteristics intimately. They're familiar with its strengths and weaknesses, its charms and pitfalls.

ARTISAN WINEMAKERS

As much as we talk about natural wine—and believe me, we talk about it a lot— wine is the product of human intervention. Apart from the viticulture itself (how and what grapes to grow) is the vinification, the process of turning those grapes into wine. The origins of winemaking stretch back to the Neolithic Era in Mesopotamia (to modern-day Georgia) and was carried on by the Canaanites and the Phoenicians, who brought large terra-cotta jars called amphorae filled with wine and ships with winemakers to the Sicilians and Sardinians. Meanwhile, the Etruscans, the forebears of today's Tuscans, had developed winemaking on their own and, in turn, taught the French everything they knew. The names of these ancient winemakers are lost to history, but today's vintners are the heirs of nearly eight thousand years of winemaking history.

The best winemakers possess vision and skill—both are necessary. Without skill, vision can never be executed; skill without vision remains merely an exercise. Think of a photographer: He must know what to capture and how to compose a shot. That is vision. But he must also know which lenses to use, the ideal exposure, the appropriate shutter speed. He must be a master of F-stops and SLRs and Adobe products. The same is true for wine. Vision is the ability of a winemaker to not just see but to perceive the terroir; skill is the possession of the not-inconsiderable chops required to portray this terroir, ushering the grape from vine to bottle in a way that preserves and amplifies its noble characteristics. This involves not just viticulture (what grapes to plant and how many vines per acre, and where and when and how to plant them) but also vinification, the process of turning the juice of grapes into wine, then aging for various amounts of time in various vessels. Every choice has the ability to bring the terroir more into focus—or not.

No one can argue that the nomadic winemaker consultants who spread the gospel of Super Tuscans to wineries throughout Italy in the 1980s and 1990s weren't skilled. And those wines, made with a combination of French varieties on Italian soil, were spectacular and highly successful on the international market. Still, the vision I admire is one that begins with the land and works organically forward, eventually reaching the market, not the other way around.

And this is, for me, the integral characteristic of an artisan winemaker: not just care for the wine in the bottle but also a regard that trellises backward to the vine and its grapes, to the land where the vines grow, to the air that rustles their leaves and the water that sustains them.

In short, I'd define an artisan winemaker as one who listens to and interprets the terroir, as expressed through its native grapes, with respect and sensitivity. To be more precise, I'd suggest the following as the key criteria artisan winemakers must meet to truly embody the term:

1. They must undertake all vineyard operations by hand. This means not using heavy machinery for plowing, planting, spraying chemicals, pruning the vines, or harvesting the grapes.

2. They must avoid the use of chemical pesticides, herbicides, and fertilizers as much as possible.

3. In the cantina (winery) itself, nothing should be added, nothing taken away. That includes additions such as cultured yeast, enzymes, and sugar, and the subtraction, through filtration, of yeasts, both live and spent, and other particle matter.

4. The winery should be independently owned and exist on a human scale. Production should be small or medium-sized. It's impossible to satisfy the standards above if production levels rise to millions of cases a year.

A Note on Natural Wine

The term "natural wine" is a slippery one that can be pushed ad absurdum. I would define it more as an ideal to which many strive than as a set of commandments engraved in stone. Certainly, making natural wine involves working organically and biodynamically if possible. Whether this is given the imprimatur of organizations like Demeter, which certifies farms biodynamic, is another matter, since many certifications are costly and, besides, many of the winemakers producing natural wines reject certification of all kinds on principle. More important than any seal is whether the winemakers follow the criteria set forth on the preceding page as best they can. I think of innovative winemakers like Lorenzo Mocchiutti of Friuli's Vignai da Duline, whose untrimmed, herbicide-free vineyards have wonderfully devolved into grape-filled meadows. Or Angiolino Maule in the Veneto, who is experimenting with all-natural fungicides in the form of seaweed and orange tonics. On the other hand, I've also heard many small, otherwise "natural," winemakers admit that given the choice between losing a vintage and the prudent use of chemical anti-fungals or something like copper sulfites, they'd choose the latter. And who can blame them—but does that then mean their wines are no longer natural? Some might say yes, but I hesitate to be so dogmatic.

Many arguments have arisen, often after the consumption of much natural wine, as to what is a natural wine and what is not. The discussion quickly moves from the use of sulfites to pondering how one can call aging in fiberglass tubs or stainless steel containers, as many winemakers do, "natural." Show me the fiberglass forests and mountains made of steel. But these are arguments, perhaps, best not had.

Like all ideals, natural wine is one that few reach completely. And natural, low-intervention methods are necessary but not sufficient in terms of making a pleasurable or worthy wine. That's why the VVV allows for subtlety and nuance. The world of natural wine, like the world itself, defies easy categorization, rebels against binary thinking, and rewards independent thought.

PART II

Emerging Styles

As the Italian wine scene has evolved, winemakers have begun looking inward and deeper at their own roots. What they've found, native grapes of stellar quality, has given them confidence and demanded from them even more technical prowess to properly express that excellence. The market has rewarded them too, embracing gutsy character-driven wines. And as this has occurred—that is to say, as we've begun to move away from the pyramid and into the world of the Vino Vero Venn—new styles have emerged, styles that are, perhaps, more suited to these evolved tastes.

Today when you walk into any decent wine store or wine bar with an Italian focus—certainly my own—you will, hopefully, find a list that goes well beyond the old red-or-white paradigm. You'll likely find an expansive section devoted to sparkling wines, and not just Prosecco, but other spumanti. You'll see rosati, a style once ignored, then maligned and mostly abandoned. Orange wines, a category that didn't even exist twenty years ago, are growing exponentially, and on some lists they rival reds. (Speaking of which, as the potential of native grapes is more fully explored, red wines are expressing themselves in various hues. And many tend lighter, with lower alcohol content and surprisingly subtle notes.) White wines, once treated like veal or castrati, are being allowed to mature and revealing themselves to be as complex as their red brethren when aged.

> Tastes cannot be separated from culture; in some ways, taste is culture.

Importantly, many of the winemakers are working organically, often biodynamically, harvesting their grapes by hand and allowing them to vinify with spontaneous yeasts in neutral vessels that don't impart overbearing notes of oak. The rallying cry for many of this crew is "Nothing added. Nothing removed."

None of this, I think, would have been possible had not a vanguard of Italian winemakers—winemakers with vision and skill—taken up the challenge to explore and champion their own terroir. Nor would this blossoming have been possible if the wine-drinking public's palate hadn't changed and broadened, providing a market for these exceptional wines. Tastes cannot be separated from culture; in some ways, taste is culture. The contemporary focus on smaller vineyards with artisanal winemakers, the impulse to eschew the easily quaffable for the actually laudable, speaks volumes about our current moment as a society. Now more than ever, we yearn for human connection, to know that there's a person on the other side of the bottle, who, like us, is from a specific place and who lives, like us, during a specific time. This is a good thing, not just because it has given rise to a proliferation of new styles (many of which aren't actually new at all), but all of which, as Luigi Veronelli says, tell the stories—*i racconti*—of those who make them.

Sparkling

Wines

Perhaps no style of Italian wine has suffered more from both a sense of inferiority and actually being inferior than sparkling wine, or *spumanti*. After World War II, during the epoch of Italian industrial wine, the few sparkling wine makers there were in Italy threw up their hands, ceding supremacy to their northwestern neighbor, France. After all, France was home to Champagne, one of the most prestigious—and heavily regulated—wines in the world. And so, instead of trying to compete, Italian winemakers went in the other direction, churning out cheap Proseccos and millions of bottles of middling, sickly sweet Asti Spumanti and the buffoonish Lambrusco.

When I first started out in wine, that was still largely the landscape. Champagne was thought of as the good stuff, worthy of esteem; Prosecco was simply fun and quaffable. But over the last twenty years, the interest in native grapes, the rising tide of the quality of Italian wines in general, and an interest in ancient methods of vinifying have caused a revolution in sparkling wine. Many producers have borrowed from the French, adopting the more handcrafted method used in Champagne (aka *méthode champenoise*, the traditional method, or, on the Italian side of the border, *metodo classico*). It's an expensive process that involves secondary fermentation in the bottle and produces a more expressive result. Others have hightailed it to the technology-free techniques of their ancestors, creating fizzy, sometimes cloudy, and often colorful wines called pet-nats (short for *pétillant naturels*, French for naturally bubbling). More accurately in Italy, these are often the product of *rifermentazione in bottiglia*, a slightly different process that yields quite similar results but without a catchy nickname.

The reason a sparkling wine sparkles is a second fermentation during which the yeasts devour the sugars in the alcohol, excreting carbon dioxide. When done under pressure, the process yields bubbles. How and when it occurs and for how long it lasts has a profound effect on the end product.

Traditionally, in the *méthode champenoise*, a still base wine is made and then blended with other individually fermented wines in a process called *assemblage*. This blend of base wines, or *cuvée*, is then transferred to bottles, where sugar and yeast are added, in a process called *tirage*. The bottles are sealed and fermented a second time, then aged on lees, the spent yeast that nevertheless endows complex flavors. Then comes *riddling*, during which the bottles are slowly turned so the lees particles and sediment travel to their necks. The necks are frozen and then the bottles opened again and the lees removed, a process called *disgorgement*. Finally, the bottles are topped off with additional wine and simple syrup, the *dosage*, and corked. It is, to put it mildly, a labor- and cost-intensive process.

Unable to compete with their French neighbors in terms of quality and market share, Italian sparkling wine producers in the *boom economico* turned instead to the *metodo Martinotti* (aka the Charmat method or tank method). This technique was invented by Federico Martinotti, a director of the Royal Oenological Station of Asti in 1895, but it was patented by the French inventor Eugène Charmat in 1907. Essentially the Martinotti method did away with the back half of the process. After the initial fermentation, bulk quantities of the base wine were fermented a second time under pressure in stainless steel tanks. The wine was then filtered and bottled. Clearly, this was a much more cost-effective process and could be done at volume. The vast majority of Italy's prodigious output of Prosecco and Lambrusco was made with this method.

> To commit to a pet-nat is a thrilling act of winemaking bravado. You don't know what you'll get until the fermentation is complete.

Affordable, sweet, and easy to drink, these sparkling wines flooded the low end of the market. The reality was, however, that sparkling wine isn't just still wine that sparkles. Though laborious, the *metodo classico* allows the yeast in the bottles to release its carbon dioxide, creating fine yet stubborn bubbles that suffuse the wine. Just as important, after the yeast has done its duty and, once exhausted, becomes lees, the lees continue to add complexity to the now sparkling wine. The Martinotti method, on the other hand, forces the carbon dioxide into the wine by sheer brute force, yielding larger bubbles but ones eager to escape. And because the wine spends less time on (fewer) lees after the second fermentation, Martinotti-produced spumanti tend to be fruit-forward but often lack depth. So anyone who could afford to drink Champagne did. Those who couldn't drank spumanti.

Happily, in the early 2000s, this state of affairs began to shift as winemakers like Vittorio Graziano, Camillo Donati, and Luciano Saetti embraced an ancient method of winemaking called *metodo ancestrale*. In this process, which had existed for centuries but had become scarce after the introduction of the Martinotti method and then nearly extinct since the 1970s, the base wine is bottled while still undergoing the last stretch of its first fermentation. Then it rests with its own naturally occurring yeasts, which, devouring the residual sugars, finish fermentation in the bottle. As the yeasts are spent, falling to the bottom of the bottle, the lees nevertheless continue to add their complexity. There is no

Grapes drying for sweet wine

assemblage. There's no *tirage*. But once the wine is bottled, the naturally occurring yeasts continue their fermentation uninterrupted until it is opened at the table.

In short order, a number of producers in the Veneto, where the process is known as *col fondo* (a term referring more to the lees at the bottom of the bottle than to the process that got them there), and Emilia-Romagna started exploring this ancestral method, a process made even more attractive by its affordability. Collectively, their output was a world removed from the crisp, clear, neat, fruit-forward but little-left-behind Prosecco of the recent past. By using the *col fondo* method, producers like Christian Zago of Ca' dei Zago, Mongarda in Valdobbiadene-Conegliano, Le Vigne di Alice, La Vigna di Iseppo, and Malibràn had a way to give their wines a voice, a voice that spoke of their land. And the cult of the ancestral method grew.

It was only a matter of time before other ambitious sparkling wine producers began applying the *metodo ancestrale* to their own wines, using the now more common term pet-nat to describe them. It's easy to see why. With so many fewer steps, making a pet-nat is much cheaper than using the *metodo classico* or the *metodo Martinotti* to produce sparkling wine, which endeared the technique to smaller winemakers of limited means.

What's more, to commit to a pet-nat is a thrilling act of winemaking bravado. You don't know what you'll get until fermentation is complete. You are at the mercy of the natural yeasts. Just as sourdough bakers rely on the power of the microbiome, so too do makers of pet-nat rely on the munificence of the world to provide for them. Pet-nat winemakers are conscientious and skilled in creating conditions for success, but once the wine is put into motion, there's no correction, no blending, nowhere to hide. One wine goes into the bottle and you stake everything you have on what emerges.

But when it does, the wines are unparalleled in their ability to express terroir. The earth has no more gifted orator. Complex, thanks to the naturally occurring yeasts, and cloudy, thanks to the often-still-present lees, pet-nats are far more complex than most spumanti I've tasted.

But although pet-nats are the new old thing, they're not the only exciting movement in Italian sparkling wine. Concurrent with the righteous valorization of native grapes, winemakers are bringing the *metodo classico* to new varieties. This itself is a different type of gamble. Instead of the yeasts, the uncertainty is in the market. At his vineyard in Piedmont, just at the Valle d'Aosta border, fifth-generation winemaker Roberto Ferrando is applying the *metodo classico* to Erbaluce di Caluso, aging the wine for thirty-six months. The results are bottles of straw-colored, Alpine-scented, invigorating sparkling wine.

My Favorite Sparkling Wines

MARTINOTTI METHOD

Sometimes it's summer and it's hot out and you want to listen to a Top 40 hit and you want to drink something chilled and bubbly and fun. Like Billy Joel's "The Stranger" or Biggie's "Juicy," there's a certain gutsy joy to sparkling wines made according to the Martinotti method. They're raw, crisp, pure, and harsh, with big swaggy bubbles and so fruit-forward it seems almost shameless. Especially when made with low-yield organically grown grapes, Martinotti-method Lambruscos and Proseccos can be the perfect summer drink.

Punta Crena, Lumassina Frizzante (Liguria); 100% Lumassina

Cieck, Erbaluce di Caluso "San Giorgio" (Piedmont); 100% Erbaluce

Sartarelli, Brut (Le Marche); 100% Verdicchio

Le Vigne di Alice, Prosecco di Valdobbiadene Brut Superiore "Doro Nature" (Veneto); 100% Glera

Casa Coste Piane, Prosecco di Valdobbiadene "Frizzante Naturalmente" (Veneto); 100% Glera

Cleto Chiarli, Lambrusco Grasparossa di Castelvetro (Emilia-Romagna); 100% Grasparossa di Castelvetro

METODO CLASSICO

Throughout Italy, winemakers are applying the *metodo classico* to their own native grapes in ways the Champenoise would never be free enough to do. From Sicily, where Marco De Bartoli makes his "Terzavia" (Third Way) from Grillo, to the Valle d'Aosta, where Ermes Pavese makes a Prié Blanc sparkling wine, the native grapes of Italy prove beyond a doubt that they deserve, and can benefit from, the labor- and cost-intensive classical method.

Ermes Pavese, Blanc de Morgex et de la Salle "Brut Zero" (Valle d'Aosta); 100% Prié Blanc

I Clivi, "R_B_L Brut Nature" (Friuli); 100% Ribolla Gialla

Ciro Picariello, Fiano Brut "Contadino" (Campania); 100% Fiano

Ferrando, Erbaluce di Caluso Spumante (Piedmont); 100% Erbaluce

Marco De Bartoli, "Terzavia" Brut Nature (Sicily); 100% Grillo

Murgo, Etna Bianco Brut (Sicily); 100% Nerello Mascalese

METODO ANCESTRALE (PET-NAT)

Pet-nats—or, more frequently but awkwardly, *rifermentazione in bottiglia* wines—are now being made in nearly half of all of Italy's regions, but they have particularly flourished, perhaps predictably, in the Veneto and Emilia-Romagna, the twinned capitals of Italian sparkling wine. With a firmly rooted culture of sparkling wine, makers in these regions are at the forefront of the pet-nat movement. In the Veneto, primary varieties include Garganega and Glera. In Emilia-Romagna, Lambrusco naturally predominates, but one finds Malvasia, Trebbiano, Pignoletto, and Ortrugo now too. Happily, these pet-nats vary from hillside to hillside, vineyard to vineyard, with different hues, individual character, and unique terroir.

Costadilà, "280 slm" (Veneto); 100% Glera

Giovanni Menti, "Roncaie Sui Lieviti" (Veneto); 100% Garganega

Monte Dall'Ora, "Vino da Sete" (Veneto); 95% Garganega, 5% Malvasia

Mirco Mariotti, "Smarazen Bianco" (Emilia-Romagna); 70% Trebbiano, 30% Malvasia di Candia

Orsi Vigneto San Vito, Pignoletto "Sui Lieviti" (Emilia-Romagna); 100% Pignoletto

Cantina Marilina, "Fedelie Rosato" (Sicily); 100% Nero d'Avola

Ca' dei Zago, Prosecco Valdobbiadene "Col Fondo" (Veneto); Glera, Verdiso, Perera, and Bianchetta Trevigiana

Mongarda, Colli Trevigiani "Col Fondo" (Veneto); Glera, Verdiso, Boschera, and Bianchetta Trevigiana

Vigneto Saetti, Lambrusco Salamino di Santa Croce "Rosso Viola" (Emilia-Romagna); 100% Lambrusco Salamino

Orange

Wines

Despite the tidiness of thinking of it that way, wine was never just red or white. The variations in hue are tremendous, from the near translucent to opaque amber. White wines actually range from pale straw to deep gold, reds from hints of pale ruby to deep purple. To frame wine as only red or white is like thinking of the sky as either bright blue or pitch black, without the brilliant display of sunrises and sunsets in between. Nevertheless, when what seems like an utterly novel color is introduced into the wine world, it's a big deal.

I remember the first time I came across an orange wine. It was 2005, and I was working at Italian Wine Merchants, where we talked about obscure bottles the way other coworkers talk about NCAA brackets or Fantasy Football leagues. A winemaker named Josko Gravner, whose small family-owned vineyard in northeastern Italy abuts Slovenia, had just released a wine he'd dubbed Breg Anfora. Starting in the 1990s, Gravner began experimenting with skin maceration, allowing the grape must to rest for various lengths of time with its skin. Skin-on maceration is common, of course. All red wines are macerated for a period of time, as are rosés, though for a much shorter period. But the skins of white grapes are rarely allowed to remain in contact with the fruit. Gravner didn't see why that needed to be the case. Just as it is for an infant, skin contact can be beneficial for a grape. As the antioxidants and tannins residing in the grape skins find their way into the wine, it develops an elegant, sturdy structure.

Like the producing of pet-nats, the making of orange wine isn't new, just the name is. According to many wine historians, the first wines discovered in Georgia were orange wines, not at all unlike what Gravner was doing eight millennia later. However, these wines, more prevalent in Georgia but also made in the bordering lands of Italy, were largely for home consumption and, like so many other styles and varieties, the making of even that orange wine—also called *vino bianco macerato,* amber wine, or skin-macerated wine—became all but extinct after World War II.

As he continued on his path to peel back the layers of time, Gravner went deeper into the soil. Literally. Not only was he allowing his grape skins partial custody with their skin, but he had also begun aging his wines in amphorae, which he buried in the land around his family's three-hundred-year-old farmhouse. The wine, made with individually fermented Chardonnay, Sauvignon, Pinot Grigio, Ribolla, and Riesling Italico, was fermented with the skins on, using only natural yeasts and without temperature control. Then it was removed and pressed, the wine poured back into the amphorae for another five months, and finished in oak casks for four years. It was bottled unfined and unfiltered.

When we poured the wine at IWM in 2005, five years after it had first been placed in the earth in those amphorae, it filled the glass with a glowing pale orange-yellow color. My colleagues and I stood gazing at it in amazement. Its texture was silky and yet still firm and the flavors that emerged—tart citrus, honey, minerals, fresh mushrooms, marmalade—were arresting.

We weren't alone in our awe. It was Gravner's "Breg Anfora" that began the recent orange wine revolution. Soon a group of fellow Friulian winemakers, including Stanko Radikon and Damijan Podversic, were bottling wines whose hues captured all the gradations of a sunset. The movement spread down the coast to Umbria, where Paolo Bea began allowing his Trebbiano Spoletino to rest for two weeks with their skins on for his "Santa Chiara" wine and even to the very tip of Sicily, where three friends—Giambattista Cilia, Giusto Occhipinti, and Cirino Strano—aged their biodynamically grown Inzolia and Grecanico skin-on must in concrete tanks.

Today we have two orange wines by the glass on offer at each of my restaurants, and I can hardly keep them in stock. Recently, orange wines have been outselling even rosati in the summer. Thanks to the pioneering efforts of Gravner, as well as some of his contemporaries, the making of orange wine has spread across Italy. But the appeal of this newly rediscovered style is international: Orange wine is being produced from Gewürztraminer in Alsace, from Garnacha in Spain, and in the Czech Republic as well as in Georgia, its ancestral home. And today you'll find orange wines in California, where winemaker George Vare planted clippings from Ribolla he had smuggled in from Friuli, and in the cellars of Steve Matthiasson of Matthiasson Wines and of Abe Schoener of the Scholium Project and Wind Gap wines.

As with any nascent development, orange wine has had its stumbles. Not every varietal is suited to the process, and not every winemaker has the patience, aptitude, or luxury to hone their expertise. Since the style is only recently rediscovered, few have experience with it. There are hangers-on too, opportunists eager to cash in on the latest trend. Like any style—and perhaps more so here, since natural wines are less forgiving (and most orange wines are natural)—errors in vinification are captured in the bottle. When an orange wine tastes more of the winemaking than it does of the fruit and the terroir—be it unpleasantly earthy or acetonic—one can't but call it a failure. Yet that is to be expected in the ferment of a popular style. I wouldn't be surprised to see orange wine being made industrially in the future. Our only defense is to be guided by our own individual tastes, asking ourselves whether and where the wine falls on the Vino Vero Venn and, most important, whether we enjoy it.

My Favorite Orange Wines

Gravner, Ribolla (Friuli); 100% Ribolla

Vodopivec, Vitovska (Friuli); 100% Vitovska

Radikon, "Jakot" (Friuli); 100% Friulano

La Biancara, Gambellara "Pico" (Veneto); 100% Garganega

Foradori, Nosiola "Fontanasanta" (Trentino); 100% Nosiola

Denavolo, "Catavela" (Emilia-Romagna); 25% Malvasia di Candia Aromatica, 25% Ortrugo, 25% Trebbiano, 25% Marsanne

La Stoppa, "Ageno" (Emilia-Romagna); 90% Malvasia, 10% Ortrugo

Camillo Donati, Malvasia Frizzante Naturale (Emilia-Romagna); 100% Malvasia di Candia

La Distesa, "Nur" (Le Marche); Trebbiano, Malvasia, and Verdicchio

Valfaccenda, Arneis "Arzigh" (Piedmont); 100% Arneis

Ezio T., Bianco (Piedmont); 100% Malvasia di Candia

Ampeleia, Bianco (Tuscany); Trebbiano, Malvasia, and Ansonica

Montesecondo, "Tïn" (Tuscany); Trebbiano, with a small amount of Malvasia

Paolo Bea, "Arboreus" (Umbria); 100% Trebbiano Spoletino

La Vallana, Bianco (Lazio); Procanico and Malvasia

L'Acino, "Chora" Calabria Bianco (Calabria); Mantonico Bianco, Guarnaccia Bianca, Pecorello, and Greco Bianco

COS, "Ramì" (Sicily); 50% Grecanico, 50% Insolia

Quality

Rosati

It will come as a surprise to absolutely no one with an Instagram account that rosé is having a moment. The style, specifically the sort of crisp, breezy rosé that hails from Provence, has become a vinous shorthand for the millennial demographic. Both share an affinity for pink and for pithy slogans in block letters on sweatshirts and tote bags. In 2018, 18.7 million cases of rosé were sold in the United States, an increase of 1.2 million cases over 2015. Yes, way, as they say, rosé.

Perhaps because the word *rosato*, the Italian equivalent of rosé, is more difficult to rhyme, thus making slogans tougher to devise, or maybe just as with Champagne, Italy had ceded to France primacy in the category, rosato, no matter the reason, was largely left out of the rosé revolution. At least internationally.

Rosato has been produced in Italy for centuries, primarily clustered in two provinces: Abruzzo, whose Cerasuolo d'Abruzzo is a deep-colored, deeply flavored rosato made with 100 percent Montepulciano, and Lombardy, near Lake Garda, where the Chiaretto style (from the Italian word for clear, *chiaro*) is made from Corvina and Rondinella grapes. However, the rosato style was not as common in Italy as its counterpart in Provence, the Loire, or the Champagne district. That began to change in the early 2000s. Not incidentally, at around the same time, many producers were allowing more maceration for their white wines to produce orange ones, while others were allowing less maceration in their red wines to make rosati. It was an era of expansion, evolution, and experimentation.

But even in the few cases of rosato production in Italy, many of the rosati were by-products. To produce the most powerful red wines possible, winemakers were using a technique called *salasso* (*saignée* in French) to decrease the ratio of juice to skin, thus creating a red wine high in flavor compounds and phenolics. The process involved allowing some of the juice of their just-crushed dark-skinned grapes to "bleed" away. The runoff, which had enjoyed only a short period of maceration, was then allowed to ferment separately, becoming a simple rosato with little depth of flavor or consideration of quality. The rosati were thus a side effect, a surprisingly fruitful side effect but merely a side effect nonetheless.

It was during my short, somewhat chaotic stint as a somm at Babbo that I first came across a bottle embodying the potential of what rosati could be when it was made with purpose. It was the summer of 2007, when elsewhere in the city, oceans of rosé were being drunk. But at Babbo, where sommelier David Lynch was careful to keep his eye trained on Italy, we offered a bottle of "il Mimo" from the Cantalupo Winery in northern Piedmont. The wine was made from 100 percent Nebbiolo grapes, grown for the express purpose of becoming rosati. Harvested earlier than one would for a still red wine, the grapes had more acidity, which yielded a deeply flavorful rosato, one reminiscent of a traditional Nebbiolo but without all the tannins and earthiness. Among wine cognoscenti, "il Mimo" was a big deal, and for a while, it was a badge of honor to have a Nebbiolo rosato on the list. Like many early entrants into the category, "il Mimo" has been somewhat eclipsed by other indigenous-grape rosati that came after it. But for me, in New York, the wine helped open my eyes to the potential of high-quality rosati from native Italian grapes.

Today, rosati are made in all twenty regions of Italy, and they highlight grapes endemic to each one. As the rosé market has grown, consumers are looking for styles that extend beyond Provence-style poolside sippers, away from wine-as-beverage and toward bottles that, through the principles of the VVV, truly express their terroir. These thirsty customers need only to look a few hundred kilometers over the Alps to Italian wines. They'll find rosati as crisp as anything Provence has to offer, but also scores of other styles, from the deep structured richness of Cerasuolo d'Abruzzo to the bright acidity and cherry flavors of Sangiovese to the distinctive smokiness of the Nerello Mascalese grown on the slopes of Mount Etna. No longer just an opening act or an afterthought, rosati are showing that, under the care of a skilled winemaker, they are capable of showcasing their terroir and are the perfect style to do so.

My Favorite Rosati

Punta Crena, "Pettirosso Allegro" (Liguria); Rossese and Cruvin

La Kiuva, Rosé de Vallée (Valle d'Aosta); Nebbiolo, Gros Vien, and Neyret

Andrea Scovero, Rosato (Piedmont); 100% Barbera

G.D. Vajra, Rosabella (Piedmont); Nebbiolo, Barbera, and Dolcetto

Nervi-Conterno, Rosato (Piedmont); 100% Nebbiolo

Pasini San Giovanni, Chiaretto (Lombardy); 65% Groppello, 25% Marzemino, 10% Barbera and Sangiovese

Togni Rebaioli, Vino Rosato (Lombardy); 100% Erbanno

Nusserhof, Lagrein Rosé (Alto Adige); 100% Lagrein

Villa Calicantus, Chiar'Otto (Veneto); Corvina, Molinara, and Rondinella

Valentini, Cerasuolo d'Abruzzo (Abruzzo); 100% Montepulciano

De Fermo, Cerasuolo d'Abruzzo "Le Cince" (Abruzzo); 100% Montepulciano

Massa Vecchia, Rosato (Tuscany); Malvasia Nera and Merlot

Fonterenza, "Rosa" (Tuscany); 100% Sangiovese

Montenidoli, Canaiuolo (Tuscany); 100% Canaiolo

Piero Riccardi Lorella Reale, "Tucuca Rosato" (Lazio); 100% Cesanese

Monte di Grazia, Rosato (Campania); Tintore and Moscio

Fatalone, "Teres" (Apulia); 100% Primitivo

I Vigneri Salvo Foti, Vinudilice (Sicily); Alicante, Grecanico Dorato, and Minnella

Camerlengo, Aglianico Rosato "Juiell" (Basilicata); 100% Aglianico

À Vita, Cirò Rosato (Calabria); 100% Gaglioppo

Frank Cornelissen, "Susucaru Rosato" (Sicily); Nerello Mascalese, Malvasia, Moscadella, and Catarratto

Fattorie Romeo del Castello, Etna Rosato "Vigorosa" (Sicily); Nerello Mascalese and Nerello Cappuccio

Age-Worthy

Whites

Sometimes you see something new; sometimes you see an old thing through new eyes. The recent rise of age-worthy Italian white wines falls firmly into the latter category. For years, wine-lover groupthink was guided by the easy-to-grasp but ultimately untrue adage that Italian red wines whose names echo through choice cellars like Barolo and Barbaresco and Brunello—the three Bs of great Italian red wine—were worth aging, while Italian white wines lacked the character and stamina to last long in the cellar. Instead, white wines were like debutantes: crisp, simple things, treasured for their freshness but whose beauty quickly faded. Like so much inherited wisdom, about both debs and wines, this was pure rubbish. Though perhaps less massive a trend than orange wine, pet-nats, and rosati, age-worthy whites are indeed gaining recognition in Italian wine circles. And as consumers recognize that many of their favorite white wines flourish in the bottle over time, winemakers have been aging them longer in the cantina as well.

The prerequisites for a wine that can withstand aging and even blossom with time are quality grapes, expert winemakers, and a market willing to pay a premium for the extra care needed to craft such a wine. For not only must the variety itself be sturdy, but it also must be grown with intention, low yields, and natural yeasts, and hand-harvested to ensure that only healthy, fully ripe fruit goes into the ferment.

The first two criteria Italy has in spades, and happily, now the third is ripening too. The world is awakening to the potential of Italian varieties like Trebbiano Abruzzese, Verdicchio, and Fiano to be world-class white wines not only upon release but also in years to come. This, of course, doesn't come as news to the myriad winemakers, like the noble family of the Boncompagni Ludovisi, who have been aging their white wines since the 1940s. It offers, however, I'm sure, satisfaction.

The estate of Tenuta di Fiorano sits between the Appian Way and the Alban Hills, just outside Rome. Since the 1940s, the patriarch of the family, Prince Alberico Boncompagni Ludovisi, had been growing Cabernet Sauvignon, Merlot, Malvasia Candida, and Semillon on the gentle slopes of the ten-hectare estate. A great friend of Luigi Veronelli, the prince was passionate about championing Italian wine made with organic methods, even as his neighbors adopted the latest agrochemical developments, and about keeping his yields low as his neighbors increased theirs. But in 1995, the prince, tired, in ill health, and apparently in bad humor as well, tore up his noble vines and retired. For four years, the future of the Tenuta was in question while its wines—"Fiorano Rosso" and "Fiorano Bianco"—aged in the estate's ancient winding cellars. The first aged Italian wine I ever drank was one of those bottles, from the last vintages under the prince. In his twilight, he had distributed his beloved bottles to true believers, one of whom was my boss Sergio Esposito at Italian Wine Merchants. Although it was early in my wine career, I had never tasted a white so aged. It was nearly as old as I was. But it wasn't just the complexity the bottle contained, though the wine was shockingly alive and energetic after all those years; it was also the poetry of the prince's story, the knowledge that I was drinking a rare bottle grown against the grain by a reclusive aristocrat with an uncompromising vision.

But you don't need to be a prince to make aged white wine. One such man who is doing so is Emidio Pepe, an elegant Abruzzian winemaker with a thin white mustache and a checkered cap. I met Emidio and his extended family at their vineyard in the hills outside Torano Nuovo. A third-generation winemaker, Emidio was one of the first to focus on age-worthy Trebbiano Abruzzese, a high-quality native grape. (He also makes a stellar Montepulciano d'Abruzzo.) From the beginning, Emidio held back some bottles of his Trebbiano, convinced the wine had the strength to withstand and reward the short-term sacrifice. The wine was deliciously mineral in its youth, and he wagered that it had strong enough bones to ensure longevity. Sitting around a table full of tomatoes from Emidio's *orto* (vegetable garden) and pasta made from wheat they grew themselves, he and his kin were like an Italian Norman Rockwell painting. I discovered, happily, that what they loved most to do was to descend into their 350,000-bottle cellar and return with a *stravecchio*—very aged—Trebbiano or their red Montepulciano, aged twenty years. Just as Prince Ludovisi's bottle of Fiorano had opened my palate to the richness of an aged white wine, so did this pale gold liquid deepen my appreciation and underline what Pepe and the prince both knew: Italian whites are worthy of age.

My Favorite
Age-Worthy Whites

Ronchi di Cialla, Cialla Bianco (Friuli); Ribolla Gialla, Verduzzo, and Picolit

Miani, Ribolla Gialla (Friuli); 100% Ribolla Gialla

Borgo del Tiglio, Friulano (Friuli); 100% Friulano

Prà, Soave Classico "Monte Grande" (Veneto); 100% Garganega

Gini, Soave "La Froscà" (Veneto); 100% Garganega

Walter Massa, Timorasso Derthona (Piedmont); 100% Timorasso

Ca' Lojera, Lugana "Annata Storica" (Veneto); 100% Verdicchio

San Lorenzo, "Il San Lorenzo" (Le Marche); 100% Verdicchio

La Distesa, "Gli Eremi" (Le Marche); 100% Verdicchio

Valentini, Trebbiano d'Abruzzo (Abruzzo); 100% Trebbiano Abruzzese

Emidio Pepe, Trebbiano d'Abruzzo (Abruzzo); 100% Trebbiano Abruzzese

Tiberio, Trebbiano d'Abruzzo "Fonte Canale" (Abruzzo); 100% Trebbiano Abruzzese

Ciro Picariello, Fiano di Avellino "906" (Campania); 100% Fiano

Benito Ferrara, Greco di Tufo "Vigna Cicogna" (Campania); 100% Greco

Deperu Holler, Isola dei Nuraghi Bianco "Prama Dorada" (Sardinia); Vermentino and Nasco

I Vigneri Salvo Foti, Etna Bianco "Palmento Caselle" (Sicily); 100% Carricante

Benanti, Etna Bianco "Pietra Marina" (Sicily); 100% Carricante

Graci, Etna Bianco "Arcurìa" (Sicily); Carricante and Catarratto

Light

Reds

For many years, the most famous Italian red wine—the Aglianico, Barolo, Barbaresco, Brunello—was a heavy, iron-clad-structured, ramrod-backboned wine bulging with tannins and bold fruit, juiced up from new French barrels. In too many cases, the innate qualities of the grape—its capacity for *tendresse* and *gentilezza*—were obliterated by the technical choices of the winemakers. But many of today's winemakers (as will become clear in the following pages) are subverting their egos in order to better listen to the grape before them and to produce lighter, low-alcohol wines free from the overbearing tannins that surged in popularity.

The best winemakers know how to liberate lightness in their variety. (Not all varieties are apt; heavily tannic grapes like Nebbiolo, Sagrantino, and Aglianico are perhaps best kept old school.) All over Italy, producers are exploring the upper registers of red wine through both choices in the cantina and a focus on lighter grape varieties, from the crunchy, crispy reds made from Cornalin in the Valle d'Aosta to the ethereal Schiava in Alto Adige to the vineyards of Pelaverga, a sort of Nebbiolo-lite from Piedmont, all the way to Vittoria, where Frappato offers pomegranate and spice but with a lightness far from what you'd expect in southern Sicily.

As tastes change, winemakers are also recasting bit players as stars. In Umbria, Ciliegiolo, a milder Sangiovese relative that's often used as a blending grape in Chianti, takes center stage in three different monovarietal wines by its greatest champion, Leonardo Bussoletti. In Sicily, the same story is unfolding in the vineyards of fourth-generation winemaker Massimiliano Calabretta, where Nerello Cappuccio, long the sideman for the more aggressive Nerello Mascalese, is answering the call for light-bodied reds.

In a cosmically ironic twist, winemakers seeking to satisfy the growing appetite for lighter red wines are challenged in their pursuit by the effects of climate change. As summers get hotter and longer, harvests need to happen earlier, to attain lower sugar and lower alcohol content in the end product. And as temperatures rise, vineyards do too, as a point of survival. The higher the vineyards, and the chillier the air, the lighter the body, so winemakers are looking for increasingly higher-altitude and east-facing vineyards. (They are also coping by reducing sun exposure, for instance, moving their vineyards from southern and western exposures to eastern and northern.) All of these efforts have had the unintended side effect of producing balanced reds that aren't too ripe, a conversation that was not happening twenty years ago.

My Favorite Light Reds

Andrea Occhipinti, "Alea Viva" (Lazio); 100% Aleatico

Dirupi, Rosso di Valtellina "Olé!" (Lombardy); 100% Chiavennasca (aka Nebbiolo)

Bussoletti, "Narni 05035" (Umbria); 100% Ciliegiolo

Grosjean, Prëmetta (Valle d'Aosta); 90% Prëmetta, 10% Cornalin

Villa Calicantus, Bardolino Classico Superiore (Veneto); Corvina, Rondinella, and Molinara

Arianna Occhipinti, "Il Frappato" (Sicily); 100% Frappato

COS, Cerasuolo di Vittoria (Sicily); 60% Nero d'Avola, 40% Frappato

Le Sincette, Garda Classico Groppello (Veneto); 100% Groppello

Calabretta, Terre Siciliane Nerello Cappuccio (Sicily); 100% Nerello Cappuccio

Burlotto, Verduno Pelaverga (Piedmont); 100% Pelaverga

Danila Pisano, Rossese di Dolceacqua (Liguria); 100% Rossese

Nusserhof, "Elda" (Alto Adige); 100% Schiava

VALLE D'AOSTA

TRENTINO-ALTO-ADIGE

FRIULI VENEZIA GIULIA

LOMBARDY

VENETO

PIEDMONT

EMILIA-ROMAGNA

LIGURIA

TUSCANY

LE MARCHE

UMBRIA

ABRUZZO

LAZIO

MOLISE

CAMPANIA

APULIA

SARDINIA

BASILICATA

CALABRIA

SICILY

PART III

Regions
& Producers

Stella di Campalto

Valle d'Aosta

Vallée d'Aoste

A small region tucked amid great peaks, sewn into the seam between Italy and France like a hidden jewel, Valle d'Aosta is one of Italy's most intriguing winemaking regions. For millennia, the sparsely populated land has been traded back and forth between tribes, empires, kingdoms, and, later, countries of Europe. Settled first by the Celts and the Ligures, the valley was overtaken by the Romans under the emperor Augustus ("Valle d'Aosta" means the Valley of Augustus), the House of Savoy, the Holy Roman Empire, and Napoleon before finally being subsumed by Italy during the Risorgimento in the 1860s. The strategic importance of the valley, cut through by the Dora Baltea River, with its mountain passes as a gateway through the Alps and into Italy, is apparent in the many castles that still stand sentinel on the steep slopes.

As each culture passed through the region, it left a trace. Today, over half of Aostians speak Italian, Aostian French, and an ancient Provençal dialect called Valdôtain. The synthesized culture, undergirded by a tough mountain mentality, is unique in Italy. Perhaps nowhere else is this flinty resourcefulness and the patina of time more apparent than among Valle d'Aosta's winemakers.

Winegrowing here is described as *viticoltura eroica,* or heroic viticulture. As one might expect for a valley hemmed in by Mont Blanc on one side and the Gran Paradiso, Italy's highest peak, on the other, there's little flatland to farm. The primary valley for wine production is the Valle Centrale, which is further subdivided into four areas: Enfer d'Arvier, Torrette, Nus, and Chambave. Like all valleys, this one is naturally broken into two slopes: the sunnier south-facing *adrét* and the less exposed *revers.* Vineyards on the *adrét* are considered more advantageous. Perhaps surprisingly for so northerly an area, the vast majority of wine produced here is red, although in recent years white wine production has increased and now accounts for 25 percent of the total. Red or white, the wines from Valle d'Aosta have pronounced acidity, a refreshing Alpine quality, moderate alcohol, and a crystalline minerality.

The vast majority of vineyards are tiny and cling to the steep slopes, chiseled into the granite and quartz and secured by ancient stone terraces. Farmers and winemakers must trudge up these impossible slopes to tend to their pergolas, harvesting the grapes by hand and returning with bins heavy with fruit.

Because most vineyards are small, many growers have historically made a little wine for their family and sold the rest of the harvest to local cooperatives that, in turn, vinify, bottle, and sell the wine from many farmers. Despite the general bias against co-op wine shared by many aficionados (those who prefer wine made from single estates), in Valle d'Aosta, many cooperatives, from the Caves de Donnas to the Cave Mont Blanc, make worthwhile wines.

Viticoltura eroica is backbreaking work, but by the end of the nineteenth century, winemaking flourished in the valley, with more than ten thousand acres under vine. Valle d'Aosta boasts a broad array of grapes. Some, like Petit Rouge, Prié Blanc, and Fumin are native, while others arrived thanks to its neighbors (and sometimes occupiers), like the Swiss grapes Petite Arvine and Cornalin, the Piedmontese Nebbiolo (here called Picotendro), and the French Pinot Noir and Gamay.

What the area lacks in size, it makes up for in the variability of its conditions. Although there's only one appellation in Valle d'Aosta, Valle d'Aosta DOC, there are seven subregions and three main valleys within it, each with a distinctive character. Warm in the summer and very cold and snowy in the winter, the vineyards vary greatly depending on their position on the mountain slopes. In the central valley, where the bulk of the grapes are grown, most vines are planted on the eastern side of the Dora Baltea, so they get the less harsh morning sunlight and are, therefore, subject to smaller diurnal shifts—that is, changes in temperature between night and day.

Famously, the region is home to some of the most elevated vineyards in the northern hemisphere, where the Prié Blanc grape grows at 1,200 meters above sea level. The soil, with the exception of the highest altitudes, is the mix of gravel and clay called glacial moraine.

Sadly, if you live in the United States, chances are you've never tasted a wine from Valle d'Aosta. In the late nineteenth century, massive outbreaks of phylloxera and plagues of oidium (aphids and fungal disease, respectively) laid waste to many vineyards. World War I and then World War II followed, laying waste to many men of vineyard-working age. In the years after the wars, Valle d'Aosta's labor-intensive winemaking didn't fit in with the hard turn to industrial production or the appetite for cheap wine, and the area under cultivation dropped precipitously, to only about 1,200 acres. But over the last twenty years, Aostian wine has been on fire. With the surging interest in native grapes, and an international palate hungry for vino vero, more exporters have been turning to the valley. Aostian wines are now found on many of the choice wine lists of the world, as they should be, and on the internet, as everything else is.

The region's labor-intensive traditions are almost by definition artisanal, and, like all things artisanal, in the midst of a comeback. A new generation of Aostians are following their grandparents' and great-grandparents' footsteps up the steep, gravelly paths to their families' vineyards and then descending with some tremendous grapes. Overgrown terraces are being reclaimed for grapes. Many of these winemakers are forgoing the traditional co-op system, instead making wines from the grapes of their own vineyards. In addition to Les Crêtes, the largest privately held winery in the region, which makes just under 20,000 cases (on the smaller side elsewhere, but huge here), there are now twelve estate growers in Valle d'Aosta.

Native Varieties

PRIÉ BLANC

In the northwestern part of the region, just south of the commune of Courmayeur, the Valdigne, or upper valley, is home to the highest vineyards in Europe, at 3,937 feet (1,200 meters). Here the gutsy, high-acid white grape Prié Blanc flourishes. It is used in some sparkling wines but most often in crisp white styles with only about 11.5 to 12 percent alcohol; it also, though more rarely, appears in a late-harvest sweet wine. While still quite a rare grape, Prié Blanc has found its champions in producers like Ermes Pavese, Maison Vevey Albert, and Maison Vevey Marziano. Pavese, for his part, makes his sparkling wine according to the *metodo classico*, but he has recently been experimenting with pet-nats as well.

PETITE ARVINE

A grape of Swiss descent and delicate temperament, at its best, Petite Arvine, grown in central Valle d'Aosta, makes for light, dry, and flinty white wines. In warmer vintages, though, Petite Arvine ripens very quickly, resulting in alcohol levels of up to 14.5 percent. The masters of Petite Arvine are Grosjean, Ottin, and Rosset, who make wines with great depth but also a streak of acidity, which prevents them from feeling heavy.

VALLE D'AOSTA

MUSCAT

A very small amount of Muscat is grown in the Chambave subzone, close to Piedmont, in the Valle Centrale. For now, there's only one producer bottling its own Muscat: Ezio Voyat, a historic producer who has made some of the greatest wines ever created in Valle d'Aosta. The winery is now run by Ezio's daughter, who spends her days as a gym teacher at a local school and her off-hours making Voyat's floral, honeyed, slightly nutty, and surprisingly dry Muscat.

CORNALIN

Cornalin is the comeback kid of Valle d'Aosta. The red grape, widely planted in the nineteenth century, was all but extinct by the 1990s. Theories for its disappearance abound, but what is certain is that the grape, which resembles Petit Rouge, only survived accidentally. Growers in Valle d'Aosta mistakenly propagated it alongside its better-known look-alike. But it has also suffered from nomenclature confusion. In Valais, Switzerland, to which the grape was exported in the twentieth century, Cornalin is called Humagne Rouge, while another grape, Rouge du Pays, is called Cornalin. For the time being, Swiss Cornalin wine is more common than the Italian. But if you do finally come across the correct Cornalin, often called Cornalin d'Aosta, you'll find a complex, elegant wine with strong tannins that carry it through—indeed, all but demand—aging of four or five years in the cellar. Happily, emerging from the edge of extinction, Cornalin, which grows best mid-slope, is now among the fastest-growing varietals in the region. Among my favorites are Grosjean's Cornalin and ViniRari Cuvée de Saint-Ours, which is a blend of Cornalin and Petit Rouge.

PETIT ROUGE

Petit Rouge, one of my favorites, is the most widely planted native red grape in the Valle d'Aosta, in the Chambave Rosso, Enfer d'Arvier, Nus Rosso, and Torrette subregions. Although small, it's a powerful grape that bears potency without weight and tempers its ripe fruit with a crisp acidity, thanks to the bitterly cold nights at altitude. Danilo Thomain is the master of the grape in Enfer. His wines are 100 percent Petit Rouge, which is rare, and the two bottlings he makes are deep, dark, and spicy—but lifted. ViniRari and Franco Noussan make delicious examples of Petit Rouge blended with Cornalin. The blend is spicy and textured, with wild berry flavors. Only Grosjean makes a varietally labeled Petit Rouge. In Chambave, the idiosyncratic Ezio Voyat (mentioned earlier) is the best producer (and the only one exported). Other arbiters no less skilled, though not in Chambave, include Didier Gerbelle and Diego Curtaz.

FUMIN

Deep and dark red, Fumin could be the long-lost sibling of Syrah. The grape is grown at lower altitudes, 1,300 to 2,000 feet, than most of the other reds in Valle d'Aosta. Although it can get a bit riper than others in the valley, it still has bright acidity and soft, juicy tannins that make its wines highly drinkable. You can find Fumin bottled by itself as Valle d'Aosta Fumin DOC, or in Torrette DOC, where it plays a minor role to Petit Rouge's lead. Try the ViniRari "Balteo," which has 85 percent Fumin and 15 percent Cornalin, or the delicious but pricey Didier Gerbelle Fumin "Ten Perdu." Also look out for Grosjean Fumin "Vigne Merletta" and Franco Noussan Vallée d'Aoste Fumin.

PICOTENDRO

The Bassa Valle, lower valley, is known primarily for two styles of wine, both of which rely heavily on Nebbiolo, known here as Picotendro or Picotener. The presence of the Nebbiolo grape makes sense, considering the region's close proximity and historical ties to Piedmont. The Arnad-Montjovet area produces a medium-bodied dry red wine made from at least 70 percent Nebbiolo, with some Dolcetto, Freisa, Neyret, Pinot Noir, and/or Vien de Nus. The area near the commune of Donnaz (or Donnas) produces wines made from at least 85 percent Nebbiolo, both of which are available only through the co-ops Caves de Donnas and La Kiuva in Arnad-Montjovet.

VIEN DE NUS

Dustier and more tannic than other Valle d'Aosta grapes, Vien de Nus is generally a blending grape, with cameos in the wines of Arnad-Montjovet, Donnas, Enfer d'Arvier, and Torrette. However, in Nus Rosso, which must be at least 60 percent Vien de Nus, the variety has its star turn. There it shows its tart cherry character with hints of dry mountain herbs, akin to an Alpine Sangiovese. For now, two producers export Nus Rosso to the States: Lo Triolet and ViniRari.

Danilo Thomain

Winemakers to Know

DANILO THOMAIN

Danilo Thomain is a man with almost demonic energy. When I visited the third-generation winemaker in May 2019, he impatiently rushed me through our tasting and hurried me along on the cellar tour, speaking in a nonstop waterfall of French. Danilo, who looks like an Italian Liev Schreiber, seemed as if he had other places he wanted to be.

I have been carrying Danilo's wines since they began to be exported to the United States in 2010. He's the only grower-producer in Enfer d'Arvier, the second-highest region here, behind Morgex et de la Salle. (The rest of the production goes to co-ops.) I am a huge fan of his wines, of which there are two, both made primarily with Petit Rouge. One is fermented in steel and aged in old oak; the other—more rustic and not exported—is both fermented and aged in even older oak barrels inherited from his grandfather. Both are fermented spontaneously, aged in non-temperature-controlled tanks, and unfiltered, and both showcase the grape's unique ability to offer heft along with a light bracing mountain verve. I had been looking forward, therefore, to meeting Danilo, but now, what was going on, I wondered. Did he want me out?

Once we were in the hot Aostian sun, though, with the Dora Baltea flowing at our backs, I understood. He pointed up a hill so steep that anywhere else it would be considered a mountain. "*Andiamo*," he said. Danilo is also, it turns out, a mountain strider. Up the hill we went, with me scrambling and huffing, Danilo gamboling as if on a pleasant stroll. By the time we reached the top, some forty minutes later, we were both dripping with sweat. But Danilo was beaming. "Turn around, Joe," he said.

I did, and I saw the entirety of the valley unfold below me. The Dora Baltea flowed through the valley floor, framed in our view by the forty-to-fifty-year-old vines planted by Danilo's father and grandfather and tirelessly cared for by him. The whole appellation is only five hectares, and the vineyards are owned either by Thomain or by members of the local co-op. It was beautiful but blazing, which accounts for the region's name, L'Enfer d'Arvier, the Inferno of Arvier, and, in part, for the demon on Danilo's label. All summer long, Danilo had been replacing the old terraces with new stone and deforesting the large trees and brush that had encroached upon them. In their place, he planted Petit Rouge vines, which were once tended here but had disappeared as the valley emptied of Aostians in mid-century, replaced, as it were, by forest. Throughout each summer, he scales the steep terraces to care for his carefully spaced vines. In October, the grapes are hand-harvested and brought down to his home, which doubles as the winery. Danilo's dedication is captured in the bottle and his energy, perhaps, in the label: a dancing demon holding a trident in one hand and a bottle of wine in the other.

ERMES PAVESE

Ermes Pavese is a first-generation winemaker whose vineyard occupies less than five acres in a small town outside Morgex, in the very far west of Valle d'Aosta, near the Swiss border and enveloped by snow-capped Alpine peaks. Ermes grows only the Prié Blanc grape, on ungrafted rootstock, at about 1,200 meters above sea level, which he'll tell you makes it the highest vineyard in Europe. The terroir offers both challenges and rewards. Thanks to the high altitude and sandy soil, phylloxera left the vines untouched, but the altitude has also caused great variability in the harvest. In 2017, Ermes lost 99 percent of his crop to hail. However, in typical Pavesian fashion, he still made a wine, Uno Percento, an orange skin-macerated Prié, whose 999 bottles each come with a handwritten note.

Today, Ermes and his wife, Milena, have been joined by their children, Nathan and Ninive. (Both have namesake wines: Nathan's is barrel-fermented; Ninive's is essentially an ice wine.) The Pavese family is devoted to exploring the potential of Prié Blanc, from a pet-nat (currently only for his own family's consumption) to *metodo classico* to fresh and crisp to oak-aged.

FRANCO NOUSSAN

Franco Noussan, an agronomy professor at a local college, and his wife, Gabriella, are the stewards of eighteen small parcels of land scattered in and around Saint-Christophe, a tiny commune on the sunnier side of the Dora Baltea. Together the pair hand-harvest all the fruit and make their wines, with a devotion to natural winemaking, in their garage turned winery. I particularly like their bottling of Torrette, a Petit Rouge–based blend with varying percentages of Fumin, Vin de Nus, Cornalin, and Mayolet. The wine is aged in stainless steel tanks and, like all Noussan's wines, naturally fermented and left unfiltered. It has terrific depth of flavor but is surprisingly bright on the palate. Their Torrette Superiore, a blend of only Petit Rouge and Fumin, is excellent as well, but riper in style.

Ermes Pavese and his son, Nathan

DAUPHIN GROSJEAN

The Grosjean family, like many Aostian clans, used to make wine solely for their own consumption. But in the 1960s, after Dauphin Grosjean exhibited his wine, made with Petit Rouge, at a local expo, the response was so encouraging that he turned his avocation into a profession. Today all of Dauphin's five sons work at what has become one of Valle d'Aosta's premier wineries. Organically certified as of 2011, it grows a mix of local grapes including Fumin, Cornalin, and Petite Rouge; French varieties like Pinot Noir and Gamay; and Swiss, including Petite Arvine. Internationally, Grosjean is probably best known for its 100 percent Petite Arvine, grown near the town of Quart. The grape's natural minerality makes it an especially promising candidate for aging.

GIULIO MORIONDO OF VINIRARI

Giulio Moriondo is a haunter of fallow vineyards, where he seeks to rescue near-extinct and forgotten varieties. Since 1987, the passionate and bookish Moriondo has been cultivating these grapes and making minuscule amounts of biodynamic, natural, unfiltered wines with them in his garage. Varieties that have Moriondo to thank for their survival are Blanc Commun and a rare white-graped mutation of Petit Rouge that Moriondo confusingly calls Petit Rouge Blanc. In addition, his ViniRari produces wines from relatively well-known grapes such as Oriou Gris, Fumin, Vien de Nus, and Nebbiolo Rose (a grape unrelated to Piedmont's Nebbiolo). Moriondo writes that his passion tires his body out but frees his soul, and, I should add, makes for some terrific and delicious wines like among others, Cuvée d'Émile, a blend of Petit Rouge and Vien de Nus.

MAURIZIO FIORANO OF CHÂTEAU FEUILLET

Most Aostian winemakers have deep roots in the region, but that's not the case with Turin-born winemaker Maurizio Fiorano, whose winery, Château Feuillet, is one of the youngest and brightest lights on the Valdostano scene. Fiorano was drawn to the land when he and his wife moved to her hometown of Saint-Pierre, where her grandfather—like so many—had maintained a small vineyard for home consumption. Starting with just 5,000 square meters in 1997, Fiorano quickly discovered he was in possession of truly exceptional terroir. The plot, on the left-hand side (*adret*) of the Dora Baltea, at 650 to 850 meters altitude, faces south, receives ideal sun exposure, and enjoys a large diurnal shift. The vineyards are in the heart of the Torrette DOC, but Fiorano grows many other native grapes as well, including Petite Arvine, Fumin, Moscato, and Cornalin.

More Exceptional Producers
in Valle d'Aosta

Caves de Donnas
Didier Gerbelle
Diego Curtaz
Ezio Voyat
Feudo di San Maurizio
La Kiuva
Les Crêtes
Lo Triolet
Maison Anselmet
Maison Vevey Albert
Maison Vevey Marziano
Nadir Cunéaz
Ottin
Piero Brunet
Rosset

Liguria

If you have, as I think you might, an image of Liguria lurking in your mind, it is probably of the colorful seaside towns of the Cinque Terre. There, clinging to the steep hills like brightly colored cubist barnacles, are the charming towns of Monterosso al Mare, Vernazza, Corniglia, Manarola, and Riomaggiore, with the astonishingly blue Mediterranean waters at their base. Perhaps this image includes dozens of the small wooden boats called *gozzi* you can almost hear bobbing in the water and, just above them, a seaside café or *friggitoria* on whose tables sit paper cones filled with *fritto misto* beside slender-stemmed glasses of golden Vermentino. But if you look beyond the foreground, if you zoom in to the blurry verdant terraced hills behind, you'll find some of Italy's most underappreciated but deeply rewarding vineyards.

For years, Liguria's wines remained largely unknown beyond the small region's borders. Some dismissed the wines—sprightly reds and crisp whites—as fit only for tourists and, in some ways, the 2.4 million visitors who tread the narrow paths between the towns annually are to blame. Much of the region's wines were drunk in those seaside cafés, and their bottles rarely made it to international shores. But there was never much wine to begin with. With 12,000 acres of vines, Liguria sits just behind Valle d'Aosta in terms of the least amount of wine produced. This sun-soaked, wind-kissed land is prime olive-growing territory and, therefore, groves, not vineyards, predominate. And, as in Valle d'Aosta, the steep terrain discourages opportunistic industrial winemakers, leaving only those both hardy and passionate.

In recent years, happily, producers have begun exporting more of their bottles internationally. Vermentino and its twin Pigato are the most common among them, but Rossese, a light, fruity, great-valued so-called *glou glou*—or easily drunk wine—from near the French border, is making its way into the world.

Much like the olives and herbs with which the grapes share the slopes of the coast, the vineyards are planted either atop cliffs, where they are buffeted by the *libeccio* wind swept north from Africa, or on stone terraces, first installed by the ancient Ligures. Liguria itself is longer than it is deep, curves along the Ligurian Sea, and benefits from the combination of seaside and mountaintop. It is divided into two subregions, the Ponente, on the western side of the Ligurian Sea, and the Levante, on the eastern. Both Ponente and Levante enjoy dramatic changes in elevation, which, though good for grapes and great for gaping, make winemaking, if not as heroic as in Valle d'Aosta, more than reasonably difficult.

Native Varieties

VERMENTINO

Vermentino, Liguria's best-known white grape, arrived sometime in the fourteenth century, brought to the region by Spaniards via Corsica. Today Liguria is the northern point of a triangle consisting in the south of Corsica and Sardinia in which the grape flourishes. Made with this exceptional Mediterranean grape, Vermentino wines tend to be fresh and lively and pair well with the seafood-focused Ligurian cuisine. Ligurian Vermentino is lauded for its herbal notes—echoes, no doubt, of the sage, rosemary, and thyme with which it shares its hills.

PIGATO

Whether or not Pigato, Liguria's *other* white, is really simply Vermentino by a different name is up for debate. Ivan Giuliani, the Ligurian winemaker behind Terenzuola, once told me that Pigato and Vermentino are like a black Labrador and a chocolate one. Pigato shares all of Vermentino's basic characteristics, but whereas Vermentino is a classic green-tinged white grape, Pigato develops a more yellowish color and often sports little brown *pighe*, or freckles, on its skin.

The two grapes enjoy a robust sibling rivalry. Often interplanted, they are hard to tell apart, but where Pigato is more widely planted, on the western side of Liguria in an area called the Riviera Ligure di Ponente, Pigato growers swear that the grape yields intensely rounder and more structured wines compared to Vermentino's more angular, saltier, and more herbal ones.

MATAÒSSU

Mataòssu—also called Lumassina or Lumaca, a name derived from the snails (*lumache*) that so often accompany the wine at the table—is a rare white grape once common in the hills of the port city Savona. Often used as part of a blend in *vino nostralini* (everyday wines), the grape ripens later than Vermentino and has long been supplanted by it. However, when vinified as a monovarietal, Mataòssu is a lively white wine with subtle aromas and low alcohol. Among the best of the grape's interpreters are the Ruffino family of Punta Crena, whose 100 percent Mataòssu called "Vigneto Reiné" pairs well not just with snails but also with anchovies, capers, and all things maritime.

ROSSESE

Of Liguria's tiny red-wine production, Rossese is king. The origins of the grape are hazy, but it is genetically linked to Tibouren, a variety used to make deep-hued rosés in the South of France. The grape grows in both the Ponente and the Levante. As a general rule, if the label says "Rossese" or "Rossese Ligure di Ponente," the wine likely contains a lighter version of the grape and is best served with a chill. If labeled Rossese di Dolceacqua, it's probably fuller-bodied and best served cellar temperature (approximately 55°F). Before the creation of the Rossese di Dolceacqua DOC in 1972, most Rossese was consumed locally, so it's a style that is still making a name for itself. These DOC wines are grown in the Ponente, where the town of Dolceacqua is located, a few kilometers from the French border. Naturally, and in this case correctly, the DOC wines are considered the highest expression of the variety. They range from light and juicy to deep and concentrated, depending on the terroir. In the fuller style, I love Maccario Dringenberg's "Posaù," Tenuta Anfosso's "Fulavin," and Danila Pisano's "Vigneto Savoia"; of the latter style, Laura Aschero and Vigneti a Prua wines are wonderful examples.

Winemakers to Know

PIERLUIGI LUGANO OF BISSON

A former art history professor, Pierluigi Lugano first came to winemaking at a preternaturally early age. When he was six years old, he helped a classmate with his family's harvest and pressing near his home in Liguria and was paid in grapes. He left his payment in a pot at his home, aping the actions of the vintner. Surprisingly, the vinification worked. However, his mother was less than enthused and punished the proud, and presumably tipsy, boy.

Over the years, his early enthusiasm has continued unabated. Lugano started as an importer of wines, but his love of native varieties led him to establish, in 1978, his own winery, Bisson. At first he only vinified, relying on well-trusted local farmers to supply both more common grapes such as Vermentino and Rossese and less common varieties like Cimixà and Bianchetta Genovese. In his tiny winery, located in the back of a *prodotti tipici* store in Chiavari (a town better known for

Tommaso Ruffino

its high-quality namesake chairs), Lugano experimented with crafting naturally fermented single-variety wines. He uses forty-eight to seventy-two hours of skin contact, which endows the wines with a rare depth, complexity, and texture. As his interest and commitment grew, Lugano decided to expand Bisson's mission to growing his own grapes.

The first wine I ever encountered from Bisson was his Bianchetta Genovese "ü Pastine," an herbal, minty, refreshing wine made with 100 percent Bianchetta Genovese, an obscure local grape. I liked it so much that whenever anyone at our restaurants requested Pinot Grigio—or Sancerre, for that matter—I would direct them instead to this bottle. When *Food & Wine* magazine's executive wine editor, Ray Isle, asked for something new, this was the bottle I delivered.

Lugano has never stopped experimenting. In 2009, he introduced a sparkling wine, also made from Bianchetta Genovese, called Abissi (the Abyss). To make it, he sends the bottles out into the Bay of Silence in Sestri to be gently rocked, under natural pressure, for eighteen to twenty-six months. When the bottles

Libera Ruffino, the matriarch of the Punta Crena family

resurface, they're bedecked with sea life and full of a brisk, lemony, and gently effervescent wine.

The process was a bit too outré for the FDA, which banned the import of Abissi to the United States, but Bisson's other wines, especially the whites, are available and worth trying. Lugano also makes a great deep-colored but light on its feet rosé from Ciliegiolo, the parent grape of Sangiovese.

THE RUFFINO FAMILY OF PUNTA CRENA

For the last five hundred years, the Ruffino family has been making wine in the Colline Savonesi in Riviera Ligure di Ponente. Their vineyards, situated on a peninsula jutting into the Ligurian Sea, are swept by sea breezes and produce an astonishing array of varieties. Today the estate is run by four siblings, including the winemaker, Tommaso, but guided by the snowy-haired Libera, who is often seen with a grandchild or two in tow.

The Ruffinos have been champions of native grapes for centuries. Today they are the sole producers of rare wines from Mataòssu and Barbarossa, a thin-skinned local red grape, not to be confused with that of the same name in Emilia-Romagna. But using traditional methods passed down through generations, Punta Crena also produces some excellent Pigato and Vermentino; Lumassina, a slow-to-ripen white; Rossese, naturally; and another long-neglected grape, Crovino.

FRANCESCA BRUNA OF BRUNA

Francesca Bruna is a true believer in Pigato. That often-overlooked grape variety, Vermentino's lesser twin, is lucky to have Bruna and her father, Riccardo, who founded the vineyard in 1970, behind it. Though a handful of producers make 100 percent Pigato wines, and a few even make two versions, only Bruna boasts a trio of them. Majé, the freshest of the wines, is made from young vines—less than fifty years old—macerated for twelve hours, and left on lees for five months in stainless steel. Le Russeghine, which uses vines that range in age from twenty-five to more than fifty years and sees a small amount of oak, has great depth and structure. "U Baccan" Pigato is a deeply structured and age-worthy wine made from a selection of grapes from vines fifty to more than seventy-five years old and aged for ten months in acacia wood. All of Bruna's wines are made from vines planted on terraced vineyards on an organic eighteen-acre mountain estate, just nine miles from the Mediterranean. I first tasted Bruna wines back at IWM in 2004, when their dedication to a relatively obscure local grape endeared the family to me. The structure, salinity, and herbaceous qualities of their wines have kept me coming back, and they've only gotten better with the years.

DANILA PISANO

Although grapevines can live for hundreds of years, they are notoriously susceptible to disease. In the late nineteenth century, phylloxera all but destroyed Italy's vineyards. Since then, mildew, mold, and black rot have been constant menaces. To safeguard and protect their vines, most winemakers use vines grafted onto more disease-resistant American rootstock. While I understand and applaud that, I can't help but get excited when I come across ungrafted vines. (It's like finding Action Comic #1 in mint condition.) That's one reason I find Danila Pisano's wines so appealing. She took over her family's small estate in 1990 and immediately began working organically. On three tiny plots of land in Liguria Ponente, totaling only 8,000 square meters, she grows the Rossese grape exclusively using a two-thousand-year-old method of training the vines called *in alberello*, whereby the vines are grown as small bushes whose leaves protect the fruit from the harsh sunlight; she has some ungrafted vines going back to 1931. (She also grows Taggiasca olives on approximately 100,000 square meters.) To drink a wine made from vines sustained only by their own roots—and, better still, fermented only with their own yeasts—is a rare experience and one of the best expressions of terroir. In recent vintages, Pisano has tilted even more toward the natural, forgoing plowing or fertilizing her vineyards in favor of planting an abundant cover crop. Her wines are distinguished by their bright, fresh character.

GIOVANNA MACCARIO OF MACCARIO DRINGENBERG

Giovanna Maccario, the current owner of Maccario Dringenberg, is the daughter of a true pioneer of Ligurian wine. In the early 1970s, Mario Maccario was bottling Rossese well before anyone else had cottoned on. When he died in 1991, Giovanna took over the vineyard and continued her father's work, relying on ungrafted vines—some dating back to the 1890s—to craft structured wines that retain Rossese's trademark juiciness.

Like Danila Pisano, Maccario trains its vines *in alberello*, and as is the case with so many Ligurian vineyards, Maccario's grapes are inaccessible to machinery, and thus they must be harvested by hand.

Giovanna believes so strongly in Rossese's ability to express terroir that she makes three separate bottlings of the variety: Classico, an approachable soft and juicy blend of different vineyard sites with high-toned fruit; "Posaù," a powerful product of a single site with alcohol levels reaching up to 14.5 percent; and "Luvaira," my favorite, another single-vineyard wine that has the depth of "Posaù" without the high alcohol content.

More Exceptional Producers
in Liguria

CINQUE TERRE

Campogrande
Ottaviano Lambruschi
Possa di Bonanini Samuele
Prima Terra
Terenzuola

COLLI DI LUNI

Il Torchio

RIVIERA LIGURE DI PONENTE

Bio Vio
Claudio Vio
Laura Aschero

ROSSESE DI DOLCEACQUA

Ka*Manciné
Tenuta Anfosso
Testalonga
Vigneti a Prua

Piedmont

Piemonte

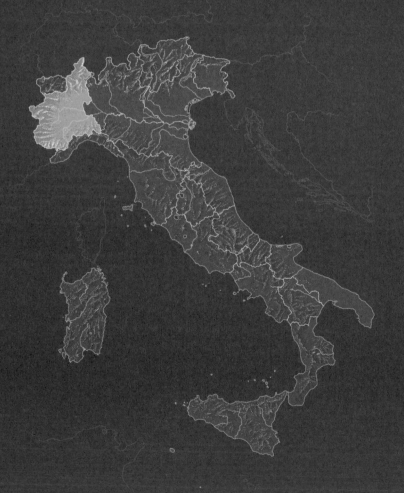

Second only to Sicily in size among Italy's regions, Piedmont is, if not the cradle, then the crown of Italian winemaking. Protected on the north by the Alps and on the south and west by the Maritime Alps, over which the sea still sends its breath, Piedmont boasts some of Italy's most highly prized—and highly priced—vineyards, those of Barolo and Barbaresco. Piedmont is red wine country for the most part, with the most commonly grown grapes being Nebbiolo, followed by Barbera and Dolcetto, but even the white wines produced here—especially Erbaluce, Timorasso, Favorita, Cortese, and Arneis—are stellar.

When wending one's way through the Po River valley in the southeastern part of the province, where Alba and Asti are the twinned centers of both Barolo and Barbaresco, it seems as if nearly every square meter of the gentle slopes is occupied by neat rows of vines. The verdant hills look combed through, like the hair of a well-groomed Grinch. With the exception of industrialized Turin, winemaking is the engine of the region's economy. Its small towns and castles, not to mention tasting rooms and *trattorie*, are full of oenophiles on pilgrimages to legendary producers like Giacomo Conterno and Bartolo Mascarello. Their passion is understandable, for in Piedmont, man and nature have conspired to create ideal conditions for viticulture. As the fog, *nebbia* in Italian, descends the mountainous slopes to hang low in the valley, it allows the thin-skinned Nebbiolo the cover to develop the aromatic complexity, tannins, and red fruit and earthy flavors that have endeared the grape to so many for so long.

Nebbiolo has been cultivated here since the thirteenth century, and Barolo has flourished since the nineteenth, championed by nobility like Juliette Colbert, a Frenchwoman married to Carlo Tancredi Falletti di Barolo, the last Marquise of Barolo, and Camillo Benso, Count of Cavour, later the first prime minister of Italy. By 1980, when both Barolo and Barbaresco were among the inaugural class of DOCG wines, the wines were already on their way to gaining an international profile. As winemakers proliferated in those zones, measurement and rules soon followed, until every square inch of the land was somehow touched by regulation. Today, simply as a measure of the attention to which Piedmontese winemaking is subject (and the high stakes involved), it's worth noting there are five individual zones, forty-two DOCs, seventeen DOCGs, dozens of crus (that is, individually bottled vineyards) within Barolo, and countless producers.

Though still a good value when compared to Burgundys of similar renown, Barolos are so expensive and the industry so well developed that most innovation and experimentation occurs outside the strictly regulated regions. But another reason for the dearth of experimentation here perhaps is that after centuries of experience, Barolo wines are as close to perfection as is possible in this world.

If the southeast of Piedmont offers the pleasure of the perfected, the north offers the joy of the unexpected. There, in Alto Piemonte, affordable acreage has lured winemakers who are now turning out some of the region's most tantalizing wines. This strip of land—along with Etna in Sicily—is the hotbed of Italian wine today. Winemakers here are using the acidic Alpine soil to make fleet-footed Nebbiolo with low alcohol and unusual delicacy, as well as whites from Erbaluce. Even in the heart of Barolo and Barbaresco, many young winemakers are taking up their family mantles and inching ever backward through time to more artisanal and natural methods, all the while stepping away from the tyranny of oak barrique.

As a young Italian wine fanatic, I'd long known and respected the wines from Piedmont, even if I couldn't afford the more rarefied bottles myself. (This was, in part, what spurred me to work in wine, where I was able to taste many wines well above my pay grade.) At the time, many of the most well-known Barolos and Barbarescos were so muscular in their youth, they were only drinkable after ten, fifteen, or twenty years of aging. By then, they were well beyond my price point as a student and, later, a struggling somm. But when I worked at Babbo, in 2007, we did a brisk trade in wines like the 1968 Conterno Monfortino and a 1989 Bartolo Mascarello Barolo among others. Regardless of the precise vintage, these bottles sent our customers into rapturous reveries. "Ah," they'd say, gazing upon a bottle of Bartolo Mascarello they had just purchased for hundreds (or even thousands) of dollars, "You haven't lived until you've seen the fog over Cannubi."

I nodded as if, of course, I'd seen the fog, but the fact was, I had never even set foot in Piedmont. I felt somewhat abashed about this fact, and I remedied it as soon as I could on my first trip there in 2008. On that first visit, I met Chiara Boschis of E. Pira Winery in Cannubi, one of the most prestigious crus of Barolo. As the first female Barolo producer and the only woman in the group of winemakers from the 1980s known as "the Barolo Boys," Chiara has been an engine toward the production of vini veri in the region. (Since then, she has evolved her style beyond the oakiness and ripeness that characterized those early innovators.) As she led us through her family's winery, she explained her plans to certify the farm as organic—a certification realized in 2014—to allow the true nature of the famed terroir to shine through by using less new oak to age her wine and leaving it both unfined and unfiltered. I marveled that even here, in the heart of the most conservative appellation, the principles of vino vero were on full display.

Native Grapes

ARNEIS

The rolling hills of Roero, in the northeast of the Cuneo province, are home to the ancient Arneis. Like so many white grapes in Piedmont, for most of history, Arneis was browbeaten by Nebbiolo. Interspersed with that variety in some vineyards, Arneis was variously used both as bait for birds, who were more attracted to its sweeter, more quickly ripening fruit than to the more expensive Nebbiolo vines, and in the vinification, to soften Nebbiolo's edges. Both practices have largely disappeared—one because of the increasing value of Arneis itself and the other because of regulation. The variety, which tends to be finicky and vulnerable to pests, demands great care and constant attention from winemakers. If left a few days too long on the vine, Arneis can yield a flat wine. But at its best, Arneis's heady combination of fruit, smoke, and white flower, cloaked in a fetching creaminess, is fueling its global popularity. Some of my favorite Arneis comes from Valfaccenda, a small valley between Canale and Cisterna, home to producers such as Vietti and Bruno Giacosa, but the biodynamic versions from Cascina Pace and Emanuele Rolfo are worth seeking out too.

TIMORASSO

In the farthest southeastern reaches of Piedmont, outside the town of Tortona, are the Colli Tortonesi. These hills are where winemaker Walter Massa worked his magic in the 1980s, single-handedly resuscitating the nearly vanished white Timorasso grape. Timorasso, also called Tinuaso or Timurasi in local dialects, was once quite common, but a combination of fussiness, asynchronous maturation (meaning the grapes ripened at different times on the same vine), vulnerability to oxidation, and the continuing adoration of Nebbiolo, drove it nearly to extinction. Massa, however, saw something in this difficult grape and, thanks to his experimentation and support, so can we. Timorasso wines are slightly tannic, honeyed, tangelo-esque, and spiced. In a region of unique wines, they are perhaps the most intriguing. Thankfully, Massa's passion project caught fire, and several other winemakers in the Colli Tortonesi, such as Claudio Mariotto, La Colombera, Roagna, and Oltretorrente, are championing the variety. In fact, when famed Barolo producer Vietti chose a second white wine to produce—the first being Arneis—it was to Timorasso that he turned.

ERBALUCE

Piedmont will always be known for its red wine, but in the northeastern corner of Alto Piemonte, in the lake district known as the Canavese, the noble white grape called Erbaluce, first introduced in the seventeenth century, grows. The grape's character is woven into its name: *erbe* means grass, and *luce* means light. Erbaluce is thus a grape eager to absorb the Alpine sun, yielding grassy and hay-like flavors in a nicely acidic minerally wine reminiscent of a Chenin Blanc. Within the Canavese, Erbaluce finds its ideal home in the valley commune of Caluso, where it is made into spumante as well as dry wine. But it is in the Passito, a sweet wine with a long history in Caluso, that I find Erbaluce most outstanding, especially those of Luigi Ferrando, whose Erbaluce wines were the first exported to the United States in 1980, and those of Cantina Favaro, a tiny producer that focuses on Erbaluce and Freisa in Caluso. Francesco Brigatti and Cieck also make Erbaluces worth seeking out.

NEBBIOLO

For centuries, Nebbiolo has been the Michael Jordan of Italian grapes. Swaggy, intense but not heavy, perseverant, the grape is virtually unstoppable in Piedmont. As the base of both Barolo and Barbaresco—both of which must be 100 percent Nebbiolo—the variety has long been the most prized, and rarely outworked by any other. Unlike the vast majority of winemaking regions in Italy, the small areas of Barolo and Barbaresco—one with 3,100 acres planted and the other with 1,200 under vine—never went through a period of *dopoguerra*, or post–World War II, bulk winemaking. Even as the rest of the country's winemakers were furiously planting and driving up their yields to satisfy the demands of the European market, winemakers like Renato Ratti, who pioneered the cru system in Barolo, identifying specific plots ideal for the cultivation of Barolo, were refining their production. Other pioneers, like Angelo Gaja, the young scion of an old winemaking family, were bottling, as early as 1967, single-vineyard wines and traveling the world preaching the gospel of Nebbiolo. Under Gaja's relentless advocacy and marketing prowess—and the work of other producers like Bartolo Mascarello, Cappellano, Giuseppe Rinaldi, and Giacomo Conterno—not only Barolo but also its hitherto underappreciated cousin Barbaresco became international sensations. Nebbiolo had become a global phenomenon, finally seen as a worthy competitor to the wine of Burgundy.

Today Nebbiolo reaches the heights of winemaking not just in Barolo and Barbaresco but in the Alto Piemonte as well, where the grape—called Spanna—

boasts a supersonic acidity and is often blended with fruitier varieties like Croatina and Vespolina, to create wines more akin to those of nearby Valle d'Aosta than the heartier reds of the Langhe.

Because Nebbiolo is known primarily through three of its most prestigious regions—Barolo, Barbaresco, and Langhe Nebbiolo—I'm going to break the format here and touch upon how the grape is expressed in each of those regions.

Barolo

Barolos are some of my favorite wines from Italy, and, in fact, the world. In this, I am not alone. These are Italy's most popular and lionized wines. But with popularity comes exposure to global trends. By the 1980s and '90s, the trend was for Barolos to get as big as possible: Nebbiolo was harvested late, macerated lightly, and aged in new French oak barrels. As mentioned earlier, this movement was led by a renegade group of men—and one woman, Chiara Boschis—nicknamed the Barolo Boys, who reveled in bucking the trends of their forebears. The results were high-octane Gordon Gekko–type wines, with what I find to be overpowering fruit and overweening levels of alcohol. (Though, to be fair, they also introduced advancements in terms of hygiene that have benefited everyone.) For the next decade or so, a schism developed between Barolo makers. On one side, "traditionalist" winemakers continued to produce less fruit-forward wines with lower alcohol levels. They relied less on new oak barriques and made wines that needed ten, fifteen, or twenty years to mature. On the other, led by the Barolo Boys, were makers who began demanding more fruit from their grapes, shortened maceration times, drew more oak from their barrels, and made wines that were ready to drink not upon release but certainly sooner than those of the traditionalists. That pendulum swung back again in the early 2000s, when the global market shifted back toward restraint. It had taken a long time, but the conversation has moved on. As the painter Yves Saint Laurent once said, "Fashions fade. Style is eternal."

Today it is not consumer trends against which Barolo makers battle, but terrifying global ones. Chief among them is climate change. As recently as the 1990s, the best vineyard sites for Nebbiolo were those where the snow melted first on the hillside—that is, the warmest plots with the most exposure to the sun. Today those once renowned *sori,* or vineyards, are scorchingly hot. Great vintages emerge only during cooler or moderate years. As a result, winemakers have crept higher and higher up the foothills of the Langhe, turned to the more sheltered vineyards facing the east, and petitioned the government to expand the borders of the DOCG region to allow for higher-altitude plantings—perhaps another reason one should be guided by taste and knowledge, not acronyms.

For those looking for the taste of Barolo without the breathtaking price tag, many producers bottle what's called Nebbiolo d'Alba, which is from an overlapping area but easier to drink young and devoid of some of Barolo's strong tannins and intensity. Other options include Langhe Nebbiolo and Langhe Rosso, especially those by producers like Burlotto and Roagna.

Barbaresco

Barbaresco is so often mentioned in tandem with Barolo (always following it) that one could be forgiven for imagining there was just one region called baroloandbarbaresco. That is profoundly not the case. Though Nebbiolo is the undisputed pound-for-pound champ of both—with Barbera and Dolcetto occupying those parcels less advantageous for its growth—Barbaresco has its own identity, born of the soil and, more saliently, from the growers upon it. A much smaller region to the northeast of Barolo, Barbaresco's terrain is similarly hilly but a bit lower than Barolo's, with a soil slightly richer in nutrients and yielding a softer, less tannic wine. Through the recent tides of the short-macerated oaky years of the 1990s to the return of long-macerated wines in larger *botti*, Barbaresco has been led by a pair of the world's best winemakers: the suave, globetrotting Angelo Gaja, who put the place on the map starting in the 1960s, and the perfectionist producer Bruno Giacosa. Meanwhile, the co-op Produttori del Barbaresco was staying true to making traditionally styled Barbaresco of very high quality. It helped perhaps that one of the world's best co-ops was making truly expressive wines right in Barbaresco for a reasonable price. With Gaja, Giocosa, and the co-op members as apostles, the gospel of Barbaresco has been heard throughout the world.

Today, as with Barolo but perhaps at a slightly quicker clip, producers are turning toward organic methodology, allowing interplanting among their vines and the interactions of wild yeasts in their cellars. Among the groundbreakers-cum-herald-bearers of Barbaresco is Luca Roagna, who grows Nebbiolo organically from old vines and produces elegant, tempered wine. I'm also particularly smitten with Teobaldo and Maria Rivella from Serafino Rivella, a pair of septuagenarians who harvest all their Nebbiolo (and Dolcetto) by hand from their prized vineyard in the cru of Montestefano. The Barbaresco of the wildman Fabio Gea has a charm all of its own.

Langhe Nebbiolo

Although the vaunted regions of Barolo and Barbaresco still have much to offer, Nebbiolo continues to find expression elsewhere. Nebbiolo wines from Langhe, the large region that contains the Barolo and Barbaresco DOCGs but also much more, offer lighter and softer iterations of Nebbiolo, with less alcohol and fewer tannins, at better value. Meanwhile, in the north, the Alpine region of Alto Piemonte is effectively the Brooklyn of Piedmont, where adventurous and upstart winemakers can afford to buy land. Perhaps the key to Nebbiolo's longevity and its primacy is the variety's ability to change with the time and place while always remaining true to itself.

BARBERA

If Nebbiolo is Jordan, Barbera is Piedmont's Scottie Pippen. Less tannic, more fruit-forward, less age-worthy, and of better value, the grape has played second fiddle to Nebbiolo for years. But don't sleep on Barbera, a wine excellent in its own right. Not only is it the most widely planted grape in Piedmont, accounting for half of all planting in the DOC regions, Barbera is capable of making some terrifically exciting wines. A versatile player, Barbera can range from light and refreshing (Fantino) to dark, concentrated, and oaky (Braida), with dozens of gradations in between. Many winemakers from Barolo are now making excellent Barberas, including G.D. Vajra, G. Conterno, Giuseppe Rinaldi, and Vietti's Scarrone Vigna Vecchia, all of whom turn out structured and earthy versions of the grape. But the home to the most expressive Barberas may still be Asti, where producers like Ezio T. and Bera make compelling natural wines thanks to low-yield farming and time-earned complexity. Also look out for San Fereolo's Langhe Rosso "Austri" from nearby Dogliani, which is often released with ten years of age and shows the age-worthiness of this grape.

DOLCETTO

The third of the Big Three Reds of Piedmont, and, I suppose, to continue with the Bulls metaphor, the Dennis Rodman of the bunch, Dolcetto is grown in Alba, Asti, and a few other regions but finds its greatest expression in Dogliani, a commune farther south, near Cuneo. While Nebbiolo can dry out your mouth and Barbera can make you pucker, Dolcetto is as dry a wine as it is sweet a grape. This sweetness is responsible for its name; *dolcetto* means little sweet one, though the wines are bone-dry. Whereas elsewhere Dolcetto is rarely afforded the respect it deserves—and therefore rarely rewards with wines worth respecting—in

Dogliani, producers such as Chionetti, Cascina Corte, and San Fereolo prove that the grape is capable of plush, well-balanced wines, whose soft fruitiness is given internal consistency and structure by subtle tannins.

However, though Dogliani continues to demonstrate its merits, the burgeoning market for Nebbiolo threatens to encroach on it. When I spoke with Nicoletta Bocca, a pioneering winemaker from San Fereolo, she expressed concern that Barolo producers were buying land here to make Langhe Nebbiolo and in the process grubbing up Dolcetto to make room for their plantings. Other producers, like Giovanna Rizzolio at Cascina delle Rose, an organic winemaker in Barbaresco, makes both an elegant and aromatic Barbaresco from her Nebbiolo and an excellent Dolcetto d'Alba. One of my favorite bottlings of this grape is the very light example from the biodynamic producer Ferdinando Principiano called "Dosset"—Dolcetto in the local dialect. It is a light, soft wine lower in alcohol than other Dolcetti.

PELAVERGA

Long simply tolerated amid the rows of Nebbiolo, the Pelaverga grape is today perhaps the sleeper hit of Piedmont. The grape arrived in the region in the seventeenth century, but it wasn't seriously cultivated until Elisa Burlotto of G.B. Burlotto convinced her father to sacrifice some of his precious Nebbiolo for a few vines of Pelaverga in 1972. After receiving DOC status in 1995, today Pelaverga is ascendant. When it's on the list at Fausto, it moves faster than a tropical storm. People are shocked and delighted to find such a light, crisp, and low-alcohol red from the heart of Barolo country. And Pelaverga continues to develop. Although it is expressed primarily as a red wine, the few producers who make it are, naturally, a daring bunch, and they have ventured into both white and sparkling Pelaverga, made according to the *metodo classico*. Along with G.B. Burlotto, Castello di Verduno, Cascina Melongis, and Fratelli Alessandria make excellent examples of this variety.

FREISA

Freisa, a strawberry-flavored grape many believe arrived with Nebbiolo in the Middle Ages, has long suffered from the comparison between the two. Although it is closely related to Nebbiolo, the hot headed Freisa boasts strong, almost untamable tannins and a swift kick of acid. In the past, the grape was used mainly in blends, relying on other grapes to mellow out its rough edges. On its own, Freisa was most often turned into either a lightly carbonated, deeply purple frizzante wine, not unlike a Lambrusco, or a sweet still wine called *amabile*, which means tender. But the expression that interests me most are the dry wines, which, though rare, reward those intrepid enough to find them. Today only about 2,500 acres of Freisa are grown, but in bottles like G.D. Vajra Langhe Freisa "Kyè," Brovia Freisa "La Villerina Secca," and Cascina 'Tavijn Vino Rosso (Freisa), the wine makes a case for its conservation.

RUCHÉ

Ruché is a lighter, aromatic red variety grown in Monferrato, northeast of Asti. Despite at times being overly aromatic and sometimes lacking in acidity, the grape comes alive in the cellars of Giuseppe Rinaldi and Crivelli, from which it emerges with its dusty floral and herbal aromas balanced with juicy fruit.

GRIGNOLINO

The Monferrato Hills roll gently between Asti and Casale, south of the Po River. Upon these slopes grow Grignolino, a local grape with a small but devoted following. So pale it's often mistaken for a rosato, Grignolino wine is surprisingly tannic, thanks to the many pips, or seeds, found within the grape; in fact, the name comes from Piedmontese dialect for "many pips." To me, these wines taste as if red cherries and oranges, peels and all, were coursed through a blender. It's a lovely wine to drink young, especially those of Francesco Rinaldi e Figli, a much-lauded producer of Barolo, and Nadia Verrua of Cascina 'Tvaijn, in Asti.

MOSCATO

Blame Drake, or applaud him, for the stratospheric rise of Moscato, here in Asti made into two sorts of sparkling wine. The Canadian rapper was among the many who name-checked the grape in the late aughts: "It's a celebration, clap clap bravo / lobster and shrimp and a glass of Moscato," he sang in 2009's "Do It Now," resulting in a steep rise in its popularity. The already prodigiously productive region of Moscato d'Asti now accounts for nearly 6 percent of all retail sales in the United States, and the grape's fragrant sweetness—present in both Moscato d'Asti and a larger-bubbled Asti Spumante—can be alluring. Unfortunately, artisanal makers of Moscato are rare. Alessandra and Gianluigi Bera, G.D. Vajra, and Vietti are the exceptions.

BRACHETTO

A rosy, lesser-known alternative to Moscato d'Asti, Brachetto d'Acqui yields a similarly sweet and lightly sparkling pink wine. At only 5.5 percent alcohol, it is light and refreshing for a dessert wine, with a certain jaunty juiciness common in frizzantes, since the bubbles brighten and lighten the wine. It was popular as a sparkling wine in the nineteenth century, but phylloxera nearly drove the grape to extinction. In the 1980s, the comeback began with winemakers near the town of Acqui Termi, including Giacomo Bologna (Braida), who perfected the expression of 100 percent Brachetto. The wine pairs well with many desserts—especially of the berries-and-cream variety—and also, unusually, with chocolate. And look out for Forteto della Luja, who makes a Passito-style Brachetto—meaning made with slightly dried grapes—and Matteo Correggia and Hilberg, who are exploring Brachetto's drier side.

CORTESE

Along with Arneis and Timorasso, Cortese makes the case that Piedmont should be known not only for its justly renowned reds but also for its white wines. For years, especially in the hurly-burly 1980s, the potential of Cortese, an eminently expressive and widely planted grape, was frittered away in industrial Gavi of middling quality. But more recently, the grape variety has attracted a small cadre of artisanal winemakers, including the biodynamic La Raia, La Mesma, and Giordano Lombardo, who are uncovering its potential to create elegant, mineral-driven, and clear wines from the hilly zone of southeastern Piedmont, where it grows best.

Fabio Alessandria

Winemakers to Know

MARIA TERESA MASCARELLO OF BARTOLO MASCARELLO

Unwavering and unparalleled, the wines of Bartolo Mascarello are standard-bearers for Barolo. The winery was founded in 1918 by Giulio Mascarello, one of the earliest examples of grower-producers, who decided to make his own wine from the family's small vineyard holdings in Cannubi, San Lorenzo, Rué, and Rocche dell'Annunziata. Mascarello reclaimed for the winery the grapes once destined for the cooperative. For years, Giulio, eventually joined by his son Bartolo, produced Barolo. The two worked together until Giulio passed away in 1981, always staying true to his original vision. In the 1960s, when individual vineyards were championed as crus, Mascarello continued to ferment and blend the fruits of his vineyards together. In the 1980s, when oak barriques came to define the style, Bartolo resisted, pasting labels reading "No Barrique. No Berlusconi" onto the crates of his increasingly prized wines. (Silvio Berlusconi was and is a popular but mind-bendingly corrupt politician who has served as Italy's prime minister three times.)

Today the vineyard is helmed by Bartolo's daughter, Maria Teresa, who remains devoted to the Mascarello way. After studying French and German in nearby Turin, she returned to the family's modest cantina in the center of Barolo to join her father in 1993. She hadn't studied winemaking, but by the time she took over the cellar four years later, she had a near century of familial experience behind her. "My school is my family," she said to me when I visited her. "The way was always artisanal."

Untroubled by trends, Maria Teresa continues to improve upon the wines: using organic methods, though not bothering with certification; hand-harvesting the family's twelve acres of grapes, which now include small plantings of Dolcetto, Barbera, Freisa, and Langhe Nebbiolo; and blending the Barolos from her four vineyards and producing some of the best wines the region has ever seen.

FABIO ALESSANDRIA OF G.B. BURLOTTO

For years, centuries really, G.B. Burlotto has been known for its light and aromatic Barolo, from the family's limestone-rich vineyard in Monvigliero. Burlotto was the official supplier to the House of Savoy, and among the many firsts the estate can boast, it was the first to sell Barolo in bottles. (Before that, it was distributed in demijohns or casks.) When G.B. died in 1927, both he and the village he had championed, Verduno, fell into obscurity. But today, thanks to his great-great-grandson, the soft-spoken Fabio Alessandria, both have taken their rightful place at the pinnacle of Barolo again. Fabio, like so many *barolisti*, or Barolo makers, sees his role as both carrying on his family's tradition and pushing it forward. "In 160 to 170 years, many things happen, no?" Fabio asks me, raising his eyebrows a bit. "Every person in my family has had a different personality and a different idea, but we always shared a classic style of wine." Fabio sees his contribution to the family business as working toward more precision and control in the winery while preserving traditions such as crushing the grapes by foot, maintaining a sixty-day period of maceration, and aging in the large wooden barrels called *botti*.

Long cherished by traditionalist Barolo fans, G.B. Burlotto has been introduced to a new demographic with their Pelaverga. Fabio tells me his family had been making wine from the grape since the 1960s but had never bothered bottling it. (Instead, they gave the wine to their friends.) When I ask Fabio if he was surprised by its popularity, he laughs and nods. "Pelaverga is living its second life. It's satisfying, because we saved the grape from extinction and now there is great interest in this variety. So, we are a little bit surprised, yes, but very happy too."

ROBERTO CONTERNO OF G. CONTERNO

Among the most storied names in Langhe is Giacomo Conterno, founder of G. Conterno, a winery in Monforte d'Alba. He established the winery shortly after World War I, before Barolo reached its vaunted status, and it achieved that status thanks in part to the punctilious winemaking of Giacomo, who professionalized his family's vinification. His flagship, Monfortino, is an elegant long-macerated Barolo, widely considered one of the world's best (with a fittingly prodigious price tag). Giacomo's sons, Aldo and Giovanni, continued the tradition until Aldo left in the 1960s to pursue a more modern style. Today Monfortino's classicism is brought to life by Giacomo's grandson, third-generation winemaker Roberto. Under Roberto's guidance, G. Conterno has expanded its vineyards into ever-more rarefied lands, including the Cerretta cru, where he produces Barolo and a powerful Barbera d'Alba. Roberto also purchased the Nervi estate in Gattinara in Alto Piemonte, where he is already making some incredible wines.

CARLOTTA AND MARTA RINALDI OF GIUSEPPE RINALDI E FIGLI

During the great Barolo wars of the 1980s and '90s, Giuseppe Rinaldi, of his namesake vineyard, was the arch-traditionalist. Rinaldi, who was known as Beppe by his friends and as Citrico, "the Acerbic One," by those who knew the acidity of his tongue, brooked no fools, nor was he impressed with the swaggering oaky, ready-to-drink Barolos of the Barolo Boys. By the time he returned from veterinary school and took over the then nearly century-old vineyard from his father, Rinaldi along with his close friend Bartolo Mascarello and a few others were holdouts for a style of long-macerated Barolos meant to be consumed years after bottling. Like Mascarello, Rinaldi refused to abide by the tyranny of crus and instead blended his Barolo, which brought him into frequent conflict with the ever-more-prescriptive bureaucracy. He was resolute about using natural yeasts to ferment and was farming organically before it became a trend.

As the pendulum has shifted away from the Barolo Boys' aesthetic over the last twenty years, Rinaldi's reputation has only grown. When I first started out, bottles of his Barolos of all vintages could be purchased year-round. Now I'm lucky to get a twelve-bottle allocation for an entire year. Giuseppe passed away just shy of his seventieth birthday in 2018. Since then, his daughters, Carlotta, the viticulturist, and Marta, the winemaker, have taken over, hewing closely to their father's insistence on making a natural, elegant, and austere Barolo. The results are just as impressive as the bottles from his time and the wines are even more hotly sought after.

CHIARA BOSCHIS OF E. PIRA

Chiara is the ninth-generation scion of the Boschis-Borgogno family, who have made Barolo in the region since 1761. With two brothers, Giorgio and Cesare, ahead of her in line to run the business, Chiara originally pursued a degree in economics and business in Turin. She only entered the world of winemaking in 1980, when a family friend, the traditionalist Luigi "Gigi" Pira, passed away without heirs, seemingly the end of a two-century dynasty. Pira's sisters turned to the Boschis-Borgogno family for help. While finishing her degree, Chiara, aided by her parents, started looking after the highly pedigreed E. Pira estates in Cannubi, San Lorenzo, and Via Nuova. She convinced her parents to buy the estate in 1981, and in 1990, she released her first wine, a Barolo Riserva Cannubi. Chiara, the only female *barolista* in the Barolo Boys cohort, began producing extravagantly audacious wines. "I was used to hearing my father say, 'My father did this. My grandfather did this. My great-grandfather did this. So we have to do this,'" she

Giuseppe Vajra

told me. "But of course, the youngest are always a little bit revolutionary. It is in our nature; we want to explore and experiment." Her experiments paid off. Her first bottling won the prestigious Tre Bicchieri award and cemented her reputation as a preeminent voice in Barolo.

Chiara was the first winemaker in Barolo I met in person, and she is still an inspiration to me. Over the years, her wines have continued to evolve, moving away from the heavy barrique and high fruit of the 1990s. Her wines are fermented using indigenous yeasts and are never fined or filtered. Her commitment to natural viticulture—including organic certification—has deepened. And her brother Giorgio has joined her, allowing the estate to expand its holdings to nearby vineyards.

GIUSEPPE VAJRA OF G.D. VAJRA

At the heart of the G.D. Vajra Winery, occupying some of the highest land in Barolo, is an unquenchable countercultural movement. The winery was founded by Giuseppe's father, Aldo, who, as a teenager, was sent to his grandparents' farm from Turin in 1968 when his father grew alarmed after he joined the widespread student protests. Aldo Vaira (the result of the 1923 Riforma Gentile meant to strip the Italian language of foreign influence; the *j* in the original name was replaced with the "more Italian" *i*) saw organic farming as a mode of resistance and, though he was new to vinification, began making wine in 1972. Occupying the middle ground between traditionalist and modernist—he was once called "the most modern of the traditionalists and the most traditional of the modernists"—Aldo forwent what he would call excessive barrique for clean Barolo aged in Slovenian barrels. Today Aldo is joined by Giuseppe, who is adept in the ways of Barolo and a proud bearer of his father's middle-way path. The Vajra family continues to cultivate their vineyards focusing on a range of Barolos as well as Freisa, Dolcetto, Moscato, and Barbera.

GIULIA NEGRI

The farmhouse atop the highest vineyard in Barolo, one of the crown jewels of Giulia Negri's estate, is called Serradenari. During the Middle Ages, when the Black Plague spread throughout the countryside, villagers would flee to the fresh air of higher ground with their money—*denarii*—tucked securely under their arms. Serradenari became known as the mountain of money. Today that fresh air buffets the Nebbiolo of Giulia Negri, the youngest and brightest of Barolo's superstars. Although the estates of nearby La Morra and Serradenari belonged to her family,

Giulia Negri

Giulia, who grew up in Rome rather than Piedmont after her parents divorced, was never interested in wine. In fact, neither was her grandfather, who tore up his vines in an attempt at truffle farming.

But in 2014, at the tender age of twenty-four, after awakening to the wonders of the vine in Burgundy, Giulia returned to Piedmont as Barolo's next superstar. She convinced her father to replant Chardonnay, Pinot Noir, and, because to not do so would be sacrilege, Nebbiolo. Soon she found herself atop the pinnacle of Barolo, standing on the slopes of Serradenari, the sea on one side and the mountains on the other.

Self-styled "The Barologirl"—a nod to the Barolo Boys of the past—Giulia has proved an immediate adept. Firmly in the artisanal mold, her Barolos are instantly recognizable for their delicate floral tones resting on surprising minerality and fine tannins. Using hand-harvesting, organic means, natural yeasts, and aging in neutral *botti*, she is both the past and the future of Barolo.

AUGUSTO CAPPELLANO OF CAPPELLANO

Teobaldo Cappellano is widely revered as the spiritual north star of the Langhe. His approach to Barolo, harvested from his family's vineyards in Serralunga, has embodied the philosophies of natural wine since the 1960s, when he took over the vineyard from his father, Francesco, who had inherited it from his father, Giuseppe, who had, in turn, inherited it from his father, Filippo. Baldo, as he was called, disdained ratings and trends with equal scorn. Not for nothing, he was also the president of the Vino Vero Association and a close friend of Luigi Veronelli. Since Baldo's death in 2009, his son, Augusto, has continued the winery's commitment to Barolo, bottled here as Barolo "Piè Rupestris," from the legendary Gabutti vineyards, and Barolo "Piè Franco," the only Barolo made from ungrafted vines, as well as a Barbera and a little-known and underappreciated digestif, Barolo Chinato.

FERDINANDO PRINCIPIANO

From his vineyards on the border between Monforte and Serralunga, Principiano lets his Nebbiolo vines intermingle with wild mint and rosemary, chamomile, and artichoke plants. Birds chirp in the trees, lured from their migratory path by the man-made pond he installed. Fruit trees like plum and pear occupy prime spots on these twenty-one hectares. Principiano's land, farmed organically since 2003, is remarkable not just for the biodiversity but also for its steep terrain—unusual for this pocket of Barolo—and for the age of the vines, some of which are octogenarians. With such natural assets, Principiano long discovered his wines needed but a light touch in the cantina, and the results, particularly his harmonious Barolo Monforte and his vivacious Dolcetto, vindicate his approach of minimal intervention.

ALDO VACCA OF PRODUTTORI DEL BARBARESCO

Produttori del Barbaresco is the rare cooperative in Piedmont that produces only Barbaresco wines and grows only the Nebbiolo grape. It was founded in 1958, when a cohort of nineteen farmers from the then-poor village of Barbaresco was convinced by a local priest to build its own winery. Among that group were both grandfathers of the current director, Aldo Vacca; his father was the director before him. Today there are fifty-four growers, all of whom are not only members of the co-op but also part owners; the co-op demands 100 percent of their Nebbiolo harvests to prevent any conflict of interest. As a whole, Produttori del Barbaresco represents 18 percent of the appellation. The wines they produce—ten Barbarescos, including nine single-vineyards, as well as some Langhe Nebbiolo—are uniformly outstanding. The members hew to the traditionalist methodology—a hand-harvest, long maceration, natural fermentation, and large oak barrels for aging—that yields Barolos cleaner in nature but essentially of the same ancient character one might have found when the cooperative started. Besides overseeing production, Vacca faces the challenging job of ensuring that all the co-op members are happy, fairly compensated—uniquely, they are paid according to the *quality* of their grapes—and incentivized not to shear off and found their own wineries. Over the course of sixty years, only five have left.

LUCA ROAGNA OF ROAGNA

The trajectory of Luca Roagna, fifth-generation heir of the Roagna Estate in Barolo, is mirrored across the winemaking regions of Italy. Born in 1981, Luca grew up inhaling the familial knowledge of his father, Alfredo, who spent his life among the Nebbiolo vines, knowledge that had been passed down to Alfredo by his own father, Giovanni, and so on through the decades. Luca went on to attend oenological school and then returned to synthesize his academic learning with age-old practices. Today, Luca is mostly found among his vines with Alfredo. As the Roagnas have always done, they harvest by hand, allow long maceration, and age in large oak barrels, falling firmly into the traditionalist style. However, Luca has introduced some refinement, including more than doubling the thickness of the *botti*, thereby allowing for a slower micro-oxygenation, which yields even more elegant wines. Thanks to Luca's work, Roagna has ascended to the upper tier of winemakers in the region. Their wines—especially the Barbaresco and, most especially, their premier bottling, Barbaresco "Crichët Pajé"—are marked by an impressive purity of fruit while never growing too overpowering. In addition to Barolo and Barbaresco, Roagna makes a stellar Langhe Rosso from vines less than twenty-five years old that could easily pass for a Barbaresco; a delicate Timorasso; Dolcetto; and even a grappa made from both Barbaresco and Barolo.

FABIO GEA

He wears a man bun. He has a room in his winery built for alien burlesque shows. The labels on his bottles feature some sort of rune halfway between the symbol for The Artist Formerly Known as Prince and a hammer and sickle. He is Fabio Gea, Barolo's most far-out winemaker. A trained geologist, Fabio returned to winemaking on this tiny, very steep plot of land bequeathed by his grandfather with little care as to what his neighbors thought. His wines can be experimental—

Fabio Gea

Nadia Verrua

he is aging his wine in handmade amphorae he calls toilets—and their names indecipherable. What, exactly, is Mushroom Panda? And they are made more inscrutable by his refusal to label them traditionally. But when it comes to producing natural wines, Fabio is unparalleled. He refuses to use toasted barrels, so as not to overpower with oak. (Instead, he uses barrels from wood cured in a volcanic-rock-heated sauna.) He neither fines nor filters his wine, and his small hand-harvested plot—which surprisingly contains not just Barbaresco but also Dolcetto d'Alba—proves that perhaps underneath the madness is rigorous method.

LUCA FACCENDA OF VALFACCENDA

Luca Faccenda occupies precisely the very center of the VVV. Born in Roero in a valley that bears his family's name, Luca, like so many other young producers, "Rumspringa'd" abroad (in his case, with another winemaker in New Zealand) before returning to his family's vineyards and starting to make his own wine in 2010. With his partner, Carolina, a graphic designer, Luca turns out bottles that are as beautiful without as they are within. It is Luca's commitment to allowing his native grapes the right of expression through minimal intervention—organic methods, natural fermentation, long maceration— that produces these wines: rare, flavorful, textured whites from Arneis, a variety whose terroir is too often obscured, and elegant, lifted reds from Nebbiolo. Together with his friends and fellow producers Enrico Cauda from Cascina Fornace and Alberto Oggero, Luca recently founded a cooperative called SoloRoero, devoted to championing their often-overlooked but never underwhelming region.

NADIA VERRUA OF CASCINA 'TAVIJN

Working from her small vineyard in Asti, Nadia Verrua—aided and abetted by her father, Ottavio, and mother, Teresa—has gradually thrown off the weighty cloak of DOC and DOCG, preferring the freedom of making wine the way she wants. Cascina 'Tavijn's Barbera is pleasingly light and fresh, but Nadia really distinguishes herself as a great interpreter of Grignolino and the lightly aromatic Ruché. Her Grignolino is called "Ottavio"; her Ruché, "Teresa"; and her Barbera, "La Bandita," her own nickname. Her wines, like their maker, are lively and full of character.

ALESSANDRA AND GIANLUIGI BERA OF VITTORIO BERA E FIGLI

The Bera family proves that it isn't the grape that falls short, driving a wine's reputation into the mud, but, too often, those who make it. Since 1964, the family has devoted themselves to Moscato d'Asti on a plot of land in Canelli, theirs since the eighteenth century, when it was bought from the Knights of Malta. They've stuck with the grape—here Moscato Bianco—through its epoch of respectability and during the long night of its massive popularity but low quality. Their vines have always been farmed organically and with a surprising amount of biodiversity. Today the winery is run by siblings Alessandra and Gianluigi, who continue the family tradition of venerating the Moscato. The vineyards are kept to a low yield, the fruit is left on the vine long enough to develop complexity of flavor, and the fermentation is low and slow. The results are a Moscato d'Asti that's lightly sweet and full of orange blossom and jasmine aromas, with a nice dryness to the finish, as well as a rare dry still Moscato. The Beras have also expanded their offerings to include a cheekily named Arcese—made from Arneis and Cortese—which is a light, bright summer sipper, and a muscular, textural, and very much alive Barbera.

NICOLETTA BOCCA OF SAN FEREOLO

What the Beras are to Moscato, Bocca is to Dolcetto: a champion. Nicoletta doesn't come from a winemaking family. She's Milanese by birth, but her father, a partisan during the war, used to bring her as a young girl to the Langhe to roam through the hills in which he'd fought, and she developed a lifelong affection for the area. In 1992, Nicoletta began acquiring the land that would become San Fereolo from a patchwork of neighbors and farmers too old or too uninterested to maintain their vineyards. Located in the western province of Cuneo, Dogliani, unlike Barolo, had not become so preoccupied with winemaking that every inch of land was cultivated and costly. Nicoletta has preserved the pockets of forest and hazelnut trees among her twelve hectares of land. And she has largely devoted herself to the careful cultivation and natural fermentation of the Dolcetto grape, historically a stand-in for Nebbiolo in the areas in which it wouldn't grow. At San Fereolo, the unusually old vines flourish under Nicoletta's ministrations. Whereas other winemakers might grub up the vines, planted between 1936 and 1978, since superannuated vines are less productive, she turns them into deeply structured wines. And unlike virtually all other Dolcetto winemakers, she allows them to rest, after natural fermentation and a long maceration, in old oak barrels for four years and in bottle for another three. They emerge, as she says, "with strength but not arrogance, . . . faithful companions, reflecting an impression of the old Piedmont as a land of treasures hidden by modesty."

Cristiano Garella is Mr. Alto Piemonte, an unstoppable one-man cheerleading squad for the region's winemakers.

Cristiano's first job out of school was at Tenuta Sella, an old estate in the once-bustling winemaking town of Lessona at the foot of the Italian Alps. In 2014, he founded his own winery with Giacomo Colombera and Colombera's father, Carlo, called Colombera & Garella.

Colombera & Garella is located in Bramaterra and Lessona, both of which boast an acidic soil that gives the Nebbiolo—here called Spanna—a nice minerality and prudent alcohol levels. The wines from Bramaterra, with its scarlet-red volcanic soil, are similarly vibrant, while those from Lessona are slightly softer but no less structured. As someone who has so tirelessly championed his terroir, Cristiano doesn't get in its way, fermenting his wines in concrete and aging them in natural oak with ambient yeasts. His deep, textural wines are truly expressions of Alto Piemonte and arguments for its return as a winemaking center.

Nicoletta Bocca

More Exceptional Producers
in Piedmont

ALBA

Hilberg

ALTO PIEMONTE

Antoniotti Odilio
Cantina Favaro
Cieck
Conti
Ferrando Vini
Francesco Brigatti
Il Sorpasso
Ioppa
Le Pianelle
Monsecco
Nervi-Conterno
Noah
Rovellotti

ASTI

Andrea Scovero
Carussin
Ezio T.
Forteto della Luja
Francesco Rinaldi e Figli

BARBARESCO

Bruno Giacosa
Bruno Rocca
Cascina Baricchi
Cascina Luisin
Cascina delle Rose
Cascina Roccalini
Gaja
Olek Bondonio
Serafino Rivella

BAROLO

Accomasso
Brovia
Cascina Melongis
Castello di Verduno
Cavallotto
Flavio Roddolo
Fratelli Alessandria
Giovanni Canonica
Giuseppe Mascarello
e Figlio
Oddero
Renato Ratti
Trediberri
Vietti

CALUSO

Cantina Favaro
Cieck

COLLI TORTONESI

Claudio Mariotto
La Colombera
Oltretorrente
Walter Massa

DOGLIANI

Cascina Corte
Chionetti

GAVI

Giordano Lombardo
La Mesma
La Raia

MONFERRATO

Braida
Cascina Gasparda
Crivelli
Iuli
Tenuta Grillo

ROERO

Alberto Oggero
Andrea Scovero
Cascina Fornace
Cascina Pace
Emanuele Rolfo

Dieter Heuskel and Peter Dipoli of Le Pianelle

Lombardy

Lombardia

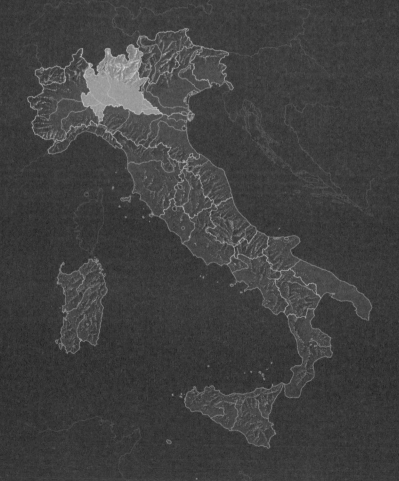

Lombardy is a large, densely populated region whose terrain extends from the Alpi Orobie in the north, flattening into the vast Padan Plain in the south, and ending at the Po River. Home to Milan, Brescia, and Lake Como, it has long been one of Italy's most industrially developed regions. The southern plains have traditionally been devoted to rice, agriculture, and cattle. (Lombardy is the ancestral home of risotto, and the diversity of its cheeses is unrivaled.) In the north, the industrial centers emptied out much of the countryside of inhabitants and workers, while the steep slopes of the Alps made widespread cultivation unappealing to all but the most hardy. The region was one of money, silk, metals, and trade. There is terrific wine to be had, but one must work hard to find it.

As can be expected from a region that includes both sub-Alpine and subtropical climates, wine in Lombardy is astonishingly diverse. Most production huddles around lakes formed ten thousand years ago as the massive glaciers traveled south. These include the Y-shaped Lake Como, Italy's deepest; its easterly twin, Lake Iseo; and, on the border between Lombardy and the Veneto, Lake Garda.

Since the 1960s, Lombardy's most productive and well-known region has been Franciacorta, on the southern shore of Lake Iseo, from whence comes a light sparkling wine of the same name. Lake Iseo and Lake Como have long served as an escape for wealthy Milanese families, some of whom, thirsty for the Champagne they enjoyed back in the city, planted French grapes that included Chardonnay, Pinot Noir, and Pinot Blanc. They began making very dry sparkling wine in the *metodo classico* from grapes grown on hillsides over which blows the cold mountain air of the nearby glacial peaks.

Of slightly more interest is a local style of Franciacorta called Satèn, which uses up to 50 percent Pinot Bianco with the balance made up of Chardonnay. It is aged on its lees for twenty-four months, rather than Franciacorta's eighteen, and is bottled at a lower pressure. The resulting wine is drier, softer, and more complex than most Franciacorta. Among the best makers of Satèn—Satènists?— are Arcari e Danesi, a pair of friends who are dedicated to championing a version of Franciacorta that truly expresses their terroir and who add no dosage at all.

To the east, on the shores of Lake Garda, one finds Lombardy's best still white wines. The lake itself serves as a natural heat reservoir, while the looming Alps protect the basin from cold northern winds. The result is an almost

Mediterranean climate, in which olive and lemon trees, caper bushes, and outlandish cacti dot the shores. Here you'll find some stellar wines, including those from a narrow band of blazingly white soil on the southern rim called Lugana. Made with the Turbiana grape, also called di Lugana, though it's really a relative of Verdicchio that evolved over time to flourish in the clay soil, Lugana has historically suffered from high yields and low quality. But when these wines are well made, as they increasingly are by producers like Pasini San Giovanni and Pratello, they can be dense without turgidity, joyful without flippancy.

On the southwestern shore, where the morainic soil also benefited from the southern path of the ancient glacier, is Valtènesi, home of a light rosato called Valtènesi Chiaretto. Like Franciacorta, Chiaretto began as an imitation of a French wine, this time a Provençal-style rosé. Here, however, it is made with Groppello, one of Lombardy's only native grapes, blended with Marzemino and Barbera. Nevertheless, this grape is often mixed with international varieties. (The DOC allows for up to 50 percent other grapes.) Somewhat confusingly, Chiaretto is also produced in the nearby Veneto, where it is called Bardolino Chiaretto. But I usually prefer the more savory, better-structured Lombardian versions, especially those from biodynamic makers like the organic-certified Pasini San Giovanni.

Moving away from the lakeshores but remaining in the Po River valley, one finds a sparkling red, Lambrusco Mantovano. Lambrusco is, perhaps, the most famous grape of Emilia-Romagna and this curiosity is the only appellation that exists outside of that region's borders. The grape, however, predates the modern borders, having been planted here since the eleventh century. Though considered by some little more than a novelty, today Lambrusco Mantovano is being made by natural wine producers like Corte Pagliare Verdieri and Fondo Bozzole intent on expressing its terroir with dignity.

Flatland is, generally, bad land for wine. Because it's so fertile, grapes don't struggle. Without struggle, they lack character. So, when plains are cultivated, they lend themselves to industrial agriculture. Thus we can travel from the southwestern border of Lombardy all the way across the Po River valley to the western border with Piedmont without missing any vines.

As one travels easterly, the ground begins to undulate, hills begin to rise, and their rolling slopes are covered in neat rows of vines. This is the Oltrepò Pavese, what some call the Tuscany of the North. The subregion is responsible for over half of Lombardy's production, most of it in the form of sparkling wine. Chief among the varieties grown here is Pinot Noir, which occupies 7,000 acres of the land and finds itself bottled primarily in sparkling wine, although, more rarely, a dry still one is also produced. But native grapes are grown here too, including the long-overlooked Croatina. Oltrepò Pavese is scattered with old winemaking estates, their cellars built on Pinot Noir. Without a doubt, though, the most interesting winemaker here is Lino Maga, who, together with his son, Luigi, tends a hillside vineyard just outside the village of Broni called Barbacarlo. Maga is a master of Croatina, and his wines are so lauded that for years his neighbors called their own wines Barbacarlo. Today he bottles both "Barbacarlo," wines made from grapes on the southwestern side of the hill, and "Montebuono" (to which Barbera is added), from the northern aspect. Both are naturally fermented, aged in old oak barrels, and turned by moonlight, bristling with character and worth the trouble to find.

What saves Lombardy from being a middling wine region with a few isolated outliers is the Valtellina, a scrunched-up area in the north, written into the south side of the Alps as if man had run out of room. As dramatic a landscape as you'll ever see, Valtellina is cut through by the Adda River, which begins at La Spedla, the highest point in Lombardy, and empties into the Po. On the north, that is, the south-facing side, the vertiginous slopes of the Alpi Retiche are steeply terraced. Because the valley is protected by the Alps and benefits from the warm *breva* breeze from Lake Como, cacti and fig trees grow here in the shadows of the mountain peaks. Just across the river on the north-facing south slope of the Alpi Orobie, dense evergreen forest grows in a truly Alpine clime.

As is the case with many appellations carved into the side of a mountain, growers vastly outstrip producers. Although there are around a thousand growers, there are only forty producers, and much of the wine is made by *négociants*, with four cooperatives accounting for the majority of the production. But, as in Valle d'Aosta, this is changing as the market expands and more growers are bottling their own grapes.

The best wine comes from five Valtellina Superiore crus: in order of descending size, Valgella, Sassella, Grumello, Inferno, and Maroggia. Though each, of course, bears its own individual style—Grumello is softer, for instance, than the more mineral Sassella or the delicate Maroggia—the Alpine Nebbiolo (here called Chiavennasca) produces a type of lighter red wine that never compromises its depth of character but isn't as intense as a Barolo can be.

Over the last century, Valtellina has seen a dramatic shift: Because the land is so difficult to farm, many growers have simply stopped farming it. Isabella Pelizzatti Perego of the winery ARPEPE told me that after World War II there were 7,500 acres of vines planted in the region; today there are only 750.

Isabella Pelizzatti Perego

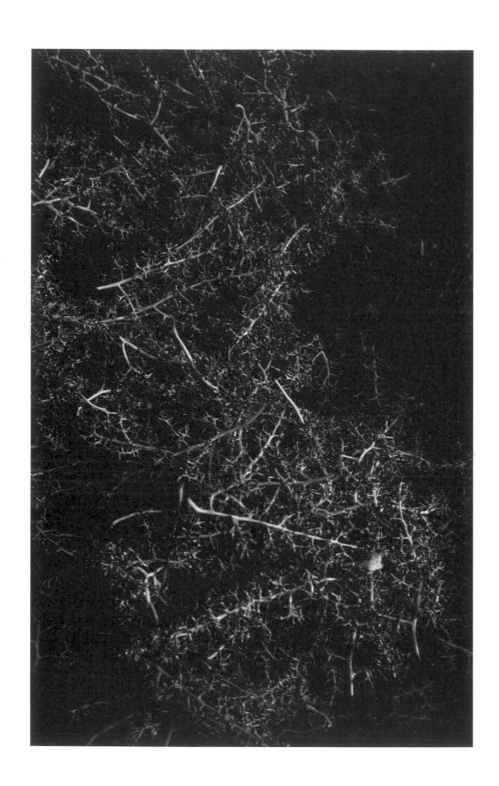

Native Grapes

NEBBIOLO

As noted, Alpine Nebbiolo, called Chiavennasca in Valtellina, benefits from the bright sun exposure and diurnal shift, yielding wines notably lighter than their Piedmontese cousins. Introduced to the region in the mid-nineteenth century, Nebbiolo almost disappeared in the postwar years as poverty, industrialization, and internal migration drained the valley of its manpower. However, in the last twenty years, producers like Isabella Pelizzatti Perego of ARPEPE and Sandro Fay have begun raising the region's profile, paving the way for even newer producers like Dirupi and, thus, greatly improving the grape's prospects.

CROATINA

Also called, confusingly, Bonarda, Croatina is a versatile red grape, high in tannins, that grows in the southwestern quadrant of Piedmont, the Oltrepò Pavese. When properly made—that is, without being overly extracted—Croatina yields a creamy, fruity wine, dark in color and with a pleasing bitterness.

GROPPELLO

Groppello is a tightly bunched red grape that grows mostly on the sandy shores of Lake Garda, where it is used in Chiaretto. (It's actually a family that includes a number of biotypes here, too similar to distinguish them from one another.) With bright acidity, notes of spice, and intense aromas, the best Groppello comes from the Valtenèsi, where it is bottled under the DOCs of Garda and Riviera del Garda Classico.

Winemakers to Know

AURELIO AND EMILIO DEL BONO OF CASA CATERINA

Aurelio and Emilio del Bono have shunned the lucrative but restrictive DOCG designation for the Franciacorta they produce from a twelve-hectare vineyard on the southeastern shores of Lake Iseo. The brothers, who learned their craft from their father, instead devote themselves to one of the purest expressions of their terroir, preferring 100 percent Pinot Noir for their *metodo classico*-made spumantes. In the fields, they use organic methods and hand-harvesting, though Aurelio, the Penn to Emilio's Teller, bucks at the word "natural" as simply a trend. Nevertheless, the brothers are ardent advocates for terroir and, fittingly, their Franciacorta is fermented by natural yeasts and left unfined and unfiltered. Seemingly unconcerned with profit, they extend their aging well beyond the standard eighteen months on lees. Though Franciacorta is traditionally drunk young, their "Cuvée 60 Nature" is aged for sixty months, while their rosato, called Antique, is aged for at least five years on the lees.

SILVIA AND FEDERICO STEFINI OF 1701 FRANCIACORTA

Another sibling pair clawing Franciacorta away from its debased reputation, Silvia and Federico Stefini founded their winery in 2012 on an eleven-hectare estate formerly owned by a count. They invested heavily in turning 1701 Franciacorta into the first certified biodynamic vineyard in the region, and their noninterventionist philosophy in the vineyard extends to the winery, where they separately ferment their holdings into a Franciacorta, a Satèn, and a rosé using indigenous yeasts and riddling by hand. Their Franciacorta is unusually dry, thanks to the lack of dosage, and while their offerings don't match the ripping acidity of Champagne, they are very well balanced and complex.

ISABELLA PELIZZATTI PEREGO OF ARPEPE

Like some sort of winemaking *Rashomon,* the genesis of ARPEPE, the finest producer of Nebbiolo in the Valtellina, is a story of revenge. The winery, originally called Pelizzatti, was founded in 1860, and for more than a century, the family tended the vineyards on the steep hills in the heart of Valtellina, stretching over the areas that have come to be known as Inferno, Grumello, and Sassella. But in 1973, a family dispute and financial pressure forced the sale of both winery and name to a Swiss conglomerate, which, in turn, merged it with another winery that was eventually sold to Gruppo Italiano Vini, a large industrial wine concern. Arturo Pelizzatti, the fourth generation of the winemaking family, disheartened that his family's tradition had met so ignominious an end, swore revenge. In 1984 he got it, purchasing back the original winery and rechristening it ARPEPE. Though he had lost the right to use his family's name, he vowed that he would restore the honor of their vines. He returned to the family's tradition of hand-harvesting and the avoidance of pesticides in the vineyard. In the cellar, he revived the family's singularly long aging process and use of large-format chestnut barrels, many of which are over fifty years old. Today the winery is run by Arturo's three children, Emanuele, Isabella, and Guido, with Isabella as the winemaker. She too has held the past in reverence while introducing innovations like a funicular that runs up and down the hillside and ever more gentle, smaller harvesting boxes.

LINO MAGA OF AZIENDA AGRICOLA BARBACARLO

Lino Maga is a little old man who has made a big name for himself through Herculean effort and stoic disregard for trends. For the last seventy years, he's been the winemaker at his family's estate, tending to the ten acres of Croatina, Uva Rara, and Ughetto on a hill called Barbacarlo, devoting himself to the cultivation of once-obscure varieties. Why? Because, as he says, it is what his ancestors did, and that is a good enough reason for him. His wine tastes not of Oltrepò Pavese, but of Barbacarlo, with its volcanic soil and southwestern exposure. And not just of Barbacarlo, but of Barbacarlo at the moment of harvest. These deep, soulful, and earthy reds capture the unbending spirit of the man who ruthlessly selects grapes at harvest, then lets them free in his winery to ferment spontaneously, without temperature control and without additives. The results are as changeable as his methods are not. Some years there's a light spritz; other vintages are completely still and all are always deeply moving.

But despite the fact that Maga is the most lauded winemaker—he's called the Bartolo Mascarello or Emidio Pepe of Lombardy—his wines are no longer DOC, labeled instead IGT Provincia di Pavia, not that this matters, or ever did, to the man himself.

ENRICO TOGNI OF TOGNI REBAIOLI

A shepherd on a mountain rising from an overlooked valley, tending to a forgotten variety: That's an accurate description of Enrico Togni, owner of Togni Rebaioli and savior of the Erbanno grape. On eight acres of steep vineyards on a slope in Val Camonica, Togni, once a law student, herds his flock of sheep and grows the difficult and tetchy local variety, which, under his care, yields a deep purple wine with great mountain-fruit character, lots of minerality, and lifting acidity. Another denomination refusenik, Togni produces a stellar San Valentino, labeled simply *vino rosso*, and a standout Erbanno rosato, among other bottles.

MIMMA VIGNOLI OF CORTE PAGLIARE VERDIERI

The fifteen hectares Corte Pagliare Verdieri occupies in the countryside near Oglio in the Po River valley seem like something lifted from a fairy tale. Beside the pumpkin patch, an auburn horse neighs as a group of laughing children feeds it oats. A long table outside is set with bottles of wine, plates of polenta, and bowls of freshly picked greens. Here, in this southeastern corner of Lombardy, Mimma Vignoli and her mother-in-law, Amedea Rossa, make an extraordinary Lambrusco Mantovano. Since Mimma has devoted only three of the fifteen hectares to cultivation of grapes (the rest is an *agriturismo* and a farm), it's hard to call the place a winery at all—but that's not a knock. The rustic vivaciousness that permeates the bottlings has as much to do with the herbs and flowers and laughter as it does with the hand-harvesting, natural yeasts, and *metodo ancestrale*. Lightly effervescent and with good acidity, they taste alive.

PAOLO AND LUCA PASINI OF PASINI SAN GIOVANNI

The evolution of a winery tends to ebb and flow. One generation establishes and the next rebels, but what of the third? At least for this current generation, it deepens the commitment to its parents' work. Such is the case at Pasini San Giovanni, makers of some tremendous Lugana and Valtènesi wines.

In 1958, Andrea Pasini founded a winery to supply his tavern on the southern shore of Lake Garda. Andrea bought his grapes from other growers for almost twenty years. Then in 1977, his sons, Bruno, Diego, and Giuseppe, bought fifteen hectares and a farmhouse in San Giovanni and began to grow their own grapes. Now the third generation, a quartet of cousins including the brothers Paolo and Luca, is doubling down on sustainability. The vineyards are now organic, the winery solar-powered, and the carbon footprint almost nonexistent. They're fully biodynamic, they hand-harvest, and they eschew new wood for their barrels. And the result is that in their hands, some lesser-known varieties—Groppello in Lugana, Turbiana in Valtènesi—show themselves capable of making a light and lively red (from the former) and a fresh elegant white (from the latter). And the Chiaretto is one of my favorite easygoing Italian rosati.

More Exceptional Producers
in Lombardy

CHIARETTO

Le Sincette

FRANCIACORTA

Arcari e Danesi
Barone Pizzini
Fattoria Mondo Antico
Il Pendio

LAMBRUSCO MONTOVANO

Fondo Bozzole

LUGANA

Pratello

OLTREPÒ PAVESE

Bruno Verdi
Castello di Stefanago

VALTELLINA

Balgera
Dirupi
Marcel Zanolari
Sandro Fay
Selva Pietro

Trentino–
Alto Adige

Südtirol. South Tyrol

Like Friuli Venezia Giulia, Trentino–Alto Adige, a small geographically varied and culturally divided area on the roof of Italy, lays bare the schizophrenia of Italy's modern borders. The southern half of the region, Trentino, belongs firmly to the peninsula below it, sharing a border and language with Lombardy and the Veneto. Alto Adige, on the other hand, which borders Switzerland to the west and Austria to the north, is as much southern Austria as it is northern Italy. It was the Treaty of Saint-Germain-en-Laye that delivered the province to Italy in 1919, a claim many South Tyroleans still regard as specious. Today the two halves of the region are largely autonomous not only from Rome but also from each other.

Binding the two regions together is the Adige River, which runs south from its headwaters in a glacial peak, cutting through a large valley. Like the rest of these northern provinces, Trentino–Alto Adige resembles the polygraph test of a poor but inveterate liar. The steep peaks of the Dolomites on the northeast and the Swiss Alps on the northwest (you can almost hear Julie Andrews warbling, "The hills are alive . . .") dominate the valley, on whose lush floor apple orchards, wildflowers, and grapevines grow.

Like neighboring Friuli Venezia Giulia and Valle d'Aosta, Trentino–Alto Adige benefits from the protection of the surrounding mountains, which both shelter and warm the valley. Winters are cold but not frigid, and in summer Trentino can feel like Palermo (but noticeably cleaner). The best vineyards in both regions are nearly vertical, clinging to the rocky walls with terrain that varies as one moves up the mountainsides. It's not unusual for small producers to harvest grapes with a 1,000-meter difference between their lowest- and highest-altitude vineyards.

But the challenges of Alpine winemaking are present here too. Most vineyards must be hand-harvested, and the backs of many winemakers are bent from the constant treks up and down the mountainsides. It is easy to see why when faced with either preserving lesser-known native varieties or planting easier-to-sell foreign ones, they chose the latter. Nevertheless, in Alto Adige, the small vineyards of about two acres each have discouraged industrial production but encouraged the hard-to-rule system of cooperatives that now produce nearly 70 percent of the wine. Many of them, including Cantina Terlano and Abbazia di Novacella, produce some wonderful single-varietals. As more care is given to these grapes and wines by the winemakers singled out in this chapter, among others, and

the market begins to warm to their efforts, a larger number of growers are doing their own vinifying.

Of native grapes, there are but a scant handful. Traditionally Alto Adige has turned its eyes elsewhere for its white grapes, including Germany for Kerner (developed only in 1929) and Sylvaner, France for Pinot Grigio, and Switzerland for Müller-Thurgau, which all grow near Bressanone. Slightly farther south, near the capital, Bolzano, you'll find red grapes, including France's Pinot Noir but also, and increasingly, the charmingly light native grape Schiava and the intriguingly dark Lagrein. Schiava, which gives a light-bodied low-alcohol red, is among the most exciting but underrated grapes in Italy.

In Trentino, best known for its sparkling white wine bottled under the Trento DOC, Chardonnay and Pinot Grigio run roughshod over the countryside, at the expense of native varieties like Nosiola and Manzoni Bianco. What red wine is made is often from Teroldego, a lively red grape championed—some would say saved—by winemakers like Elisabetta Foradori at Foradori.

Overall, the region is mid-stride in its vino vero evolution. The back half of the twentieth century wasn't all that hospitable to the production of high-quality wine here. In the 1960s, the German market for bulk wine drove producers in Trentino to favor quantity over quality. A thirst for white wines in the '70s led to the widespread abandonment of Schiava, the region's signal red grape. Even as the pendulum began to swing back toward independent growers, organic viticulture, and native grapes, because of the extensive use of the co-op system, change has been slow in coming. And the deep roots of foreign grapes have made it difficult for native varieties to take hold. But between Teroldego, its genetic offspring Schiava and Marzemino, Lagrein, and Nosiola, this strange dual region is finding itself. Independent winemakers are making their voices heard and today the native grapes themselves, singing of their mountainous and unique terroir, are raised in chorus. The hills are alive indeed.

Native Grapes

NOSIOLA

True to its name, the Valle dei Laghi is dotted with dozens of vibrant blue and aquamarine lakes, made all the more striking by the terraced hills and distant mountain peaks. The Valle dei Laghi is home to Nosiola, one of Trentino's only native white grapes. Though long grown in these hills, Nosiola was most frequently blended with other grapes or used in a sweet Vin Santo. (In 1930, the Archbishop of Trento decreed that only Nosiola wines be used during the Holy Sacrament, so as not to stain the priestly garments.) But when vinified as a dry white wine, Nosiola yields a citrusy and refreshing white, with a signature nuttiness and faint salty character. (This is especially true of Nosiola from Cesconi, nestled into the hills of Pressano.)

Only recently has the grape's capacity been explored by winemakers like third-generation Francesco Salvetta, who organically farms four acres of Nosiola on three separate cliffs tucked between the limestone crags of the Dolomites and the spring-fed Sarca River, and the pioneering Elisabetta Foradori. Most exciting is how experimental and free these winemakers are, as if eager to make up for the long neglect under which Nosiola suffered. Both make an orange skin-macerated Nosiola whose pale color belies its nutty flavor and creamy character.

TEROLDEGO

By the 1970s, this once popular and praised red grape had become so rare it took one determined woman, Elisabetta Foradori, to nurse it back to life. Now, thanks in large part to her efforts, and her encouragement of other winemakers like Zeni, Fedrizzi Cipriano, and Nusserhof to grow it, Teroldego is once again the most important red grape of Trentino. A cousin of Syrah and father of Alto Adige's native Marzemino and Lagrein, Teroldego yields light, lively reds that are nevertheless deep-colored and darkly fruited and age well. It's this unusual ability that flags Teroldego as a breakout grape. It's grown throughout Trentino (and now Tuscany, Sicily, the United States, and Australia), but the best Teroldego in the world hails from the flood plain north of Trento, the Piana Rotaliana.

SCHIAVA

As a parable whose lesson is that time is circular, one need only look as far as Schiava, an ancient red grape native to Alto Adige. For centuries, Schiava commanded both respect and a premium price. Records of wills from the fourteenth century mention it; receipts from the sixteenth century show that Trentinesi were willing to pay more for Schiava than for other grapes. Its name means slave, referring to the way the vines were traditionally lashed to poles during cultivation. References from the Middle Ages to Schiava—also called Vernatsch in Alto Adige—abound. But by the early 2000s, Schiava was in a diminished state. The market had turned its back on these very pale red wines with low alcohol, few tannins, and little to no acidity. And so what acres of Schiava remained there were left to shrivel.

Now, the appeal of an unassuming, undemanding, but rewarding light red wine has again demanded our attention. And more producers are once again growing and vinifying Schiava. From Galea, whose century-old vines are grown by Nals Margreid, to Glögglhof, Franz Gojer's estate with superannuated vines grown at high elevations, Schiava seems to now be the future of Alto Aldige.

LAGREIN

Difficult to grow, difficult to make, Lagrein nevertheless has played an important part in Alto Adige's red-wine landscape. Historically made into a deep-colored still wine, it was traditionally reserved for nobles and clergy, which gave rise, in part, to the German Peasants' War in the early 1500s. (Among other demands, the peasants wanted their Lagrein.) The grapes are often cursed with sterile flowers and even in the best of times have little pollen. They ripen late and are thus exposed longer on the vine. It takes patience, a steely will, and, perhaps, a bit of masochism to devote oneself to Lagrein, but, happily, Alto Adige seems full of winemakers who fit the bill. One of today's most exciting expressions of Lagrein is a superb complex rosato labeled either as such or as Lagrein Kretzer. These wines range from fruit-forward rosés from Muri-Gries to the clean and peppery rosés of Nusserhof in Bolzano. As a still wine, Lagrein can be opaque, heavily tannic, and overbearing. But in the right hands, like those of Nusserhof or Manincor or Ansitz Waldgries, those tendencies are tamed and a charming *gentilezza* emerges.

MARZEMINO

Though present in Lombardy as a blender (see page 123), in Trentino, Marzemino is more often made into a 100 percent varietal with grassy and herbal notes, a bright acidity, and more balance than a gyroscope. The best, grown on the banks of the Adige River, hail from Eugenio Rosi.

Elisabetta Foradori

Winemakers to Know

When I first opened Fausto, I commissioned three black-and-white line-drawn portraits to hang on one wall of the dining room. These would be our guardian spirits. On the left is Emidio Pepe, patriarch of the Pepe family winery in Abruzzo, maker of Trebbiano d'Abruzzo, Montepulciano, Pecorino, and a rare rosato. On the right is Arianna Occhipinti, the young iconoclastic Sicilian winemaker and champion of Frappato. And in the middle is Trentino's Elisabetta Foradori, the savior of Teroldego. Of all the Italian winemakers I could have chosen, these three pioneers—each with their own vision, belonging to different generations, with their hearts buried in the soil of their disparate homes—occupy the exact center of the Vino Vero Venn. To me they are the apotheosis of Italian winemaking, and I resolved from the start to pour only wines that could pass their (unseeing) gaze.

The Azienda Agricola Foradori sits in the middle of the village of Mezzolombardo, with the Monte di Mezzocorona on one side and the Dolimiti di Brenta on the other. Purchased by Elisabetta's grandfather in the 1930s, the estate, three houses amid rows of vines, was passed to her father, who died unexpectedly in 1976, when Elisabetta was twelve years old. Her mother ran the place—a few acres of an obscure and little-respected grape, Teroldego—as regent until Elisabetta turned nineteen, at which point she took to the vines with a righteous fury, deep indignation, and what would become a lifelong commitment to the variety.

At the time, Trentino's vineyards were under assault from international varieties. Their local grapes, the grapes her grandfather had planted and her father tended, grapes like Teroldego, Nosiola, and Manzoni Bianco, were thought incapable of producing quality wines. Intent on proving them wrong, Elisabetta began coaxing her vines toward production. Using mass selection (*sélection massale* in French)—that is, actually planting as opposed to hand-selecting clones—she expanded the vineyards. Trusting in the sandy alluvial soil of the Piana Rotaliano, she steered the farm toward working biodynamically and organically, allowing flowers and vegetables to grow among the vines. Most of all, she let the grapes do the talking. The first wines, Foradori Vigneti Delle Dolomiti IGT, were a hit, but Elisabetta, the "Queen of Teroldego," wasn't satisfied. The wines still bore too much oak, she thought, thanks to her use of new French barrels, and had lost their connection to the land. So she jettisoned those barrels and, starting in 2000, began a period of radical experimentation. Now she uses a mixture of cement, stainless steel, and acacia barrels and amphorae. She shows equal respect to her other varieties, including Nosiola and Manzoni Bianco. The former is macerated for seven to nine months in clay amphorae, the latter in neutral acacia wood. The response to her wines was immediate, and her belief was vindicated.

Today, Elisabetta has passed the day-to-day operation of the cantina to her three children: Emilio, who is now the winemaker; Theo, who runs the business side; and Myrtha, a farmer who has pushed Foradori ever closer to becoming a farm. Today crops of herbs and beans grow among the Teroldego vines.

As for Elisabetta, she continues to inspire Italian winemakers, returning again and again to her roots and refusing to rest on her laurels. And she's never wavered in her mission. Recently she purchased vineyards even higher in the mountains to plant nearly extinct indigenous grapes like Vernaccia Trentina and Verdealbara, convinced that with a hand sufficiently skilled and an ear acutely attuned, these overlooked local varieties will be as promising as Teroldego.

HEINRICH AND GLORIA MAYR OF NUSSERHOF

An accidental urban winery, Nusserhof is hidden behind a white gate nearly in the center of Bolzano. But in 1788, when the estate was established on the banks of the Isarco River along a line of walnut trees—*nusser* means nut—this was still the countryside. Today it's an oasis, with Eida and Heinrich Mayr and their daughter, Gloria, maintaining a proper English garden filled with rosebushes and fruit trees, playing music together in the parlor, and making some of Alto Adige's most erudite wines.

Despite attempts over the years to uproot the house and its inhabitants, Nusserhof stubbornly remains. One reason, as Gloria tells me, is that her uncle, a member of the Resistance to the Nazi occupation who died in a concentration camp, was beatified in 2017 by Pope Francis. This is holy land in a sense, and so the family holds out.

It is perhaps no surprise that their steadfastness to a Bolzano of old leads them to champion native grapes such as Lagrein, Teroldego, and Blatterle, an extremely rare white variety. These grow in the tiny 2.5-hectare plot surrounding the house. On a hillside nearby, on an even smaller plot, grow a few Schiava vines.

Of Lagrein, the Mayrs make two: a rosato full of crushed fresh fruit and black pepper, and a Riserva, which, after five years, reveals itself as both vibrant and brooding. Their tank-aged Blatterle, aged with its skins for two weeks, smells like elderflowers and has the tartness of a great McIntosh apple. And the Schiava is unusually robust and aromatic. All are fermented naturally, unfiltered, and unfined.

EUGENIO ROSI OF PERCISO

Along with Elisabetta Foradori and eight other winemakers, Eugenio Rosi is a founding member of I Dolomitici, a group of winemakers based in Mezzolombardo and environs whose manifesto closely aligns with the VVV. "When healthy wine-growing allows the vine its full expression," they wrote, "the fruit will express its authenticity; consequently, technology in the winery (and the standardized tastes that it leads to) will become useless." Amen.

For Eugenio, the fruit in question is primarily a rare strain of Lambrusco called Lambrusco a Foglia Frastagliata after its jagged leaves, and of which only twenty hectares are under vine in Trentino. His vineyard, PerCiso, is named for an old Trentino farmer, Narciso, who tended these century-old vines and donated them to Eugenio when he started the winery in 1997. Rosi vinifies not only Lambrusco but also an orange Nosiola from the cantina of Elisabetta Foradori.

MATTEO FURLANI OF CANTINA FURLANI

You'd be forgiven for doing a double take if you passed Matteo Furlani's cantina in Povo, the village just east of Trento, in the winter. You're likely to find large glass demijohns nestled neatly in the snow. They are all part of fourth-generation winemaker Matteo's ongoing project of showcasing native grapes in natural ways. In his high-altitude vineyards, the young Furlani grows Nosiola and Marzemino, as well as even rarer varieties like Joannizza, Lagarino, Negrara, and Verdealbara. He trains the vines using the *pergola Trentino* method, a semi-protected system that offers good exposure and ventilation; never uses chemicals; and harvests by hand.

In the cantina, Matteo's interest lies mainly with making sparkling wines, a form of expression rarely afforded to these varieties. Intent on preserving their unique character, he prefers skin-on maceration and allowing the wine's natural yeasts to spark secondary fermentation in the bottle, the *sui lieviti* method. Not one to be hamstrung by tradition or trends, Matteo also grows small amounts of Chardonnay and Pinot Noir, which he turns into a delicate spumante using a truncated variant of the *metodo classico* called *medoto interotto*, in which the fermentation is arrested (*interotto* means interrupted). And instead of filtering, he prefers to place his wines in said demijohns in the snow in winter to allow the sediment to settle out.

The output of this small winery—bottles ranging in hue from the bubblegum pink of his macerated *sui lieviti* to the dark Mino Rosso, a still red made from Marzemino—showcases the potent and versatile possibility of Trentino's terroir.

BRIGITTE AND PETER PLIGER OF KUENHOF

Just outside the city of Bressanone in the Valle Isarco, a former Zen student named Peter Pliger; his wife, Brigitte; and now their son, Simon, tend six hectares of small, narrow terraces high above the valley floor. Based in a twelfth-century farmhouse that once belonged to the Bishop of Bressanone, the Pligers grow a slew of grapes, including Gewürztraminer, Riesling, Sylvaner, and Grüner Veltliner, all of which do well in the austere schist-and-quartzite soil and under Pliger's organic attitude. (However, he has no interest in pursuing certification.) In an area dominated by the *négociant* system, Pliger was one of the first winemakers to vinify his own grapes, when he founded Kuenhof in the 1990s, as well as one of the first in the valley to plant Riesling. Both his Rieslings, especially "Kaiton," and other varietal bottlings display a bone-dry nature yet burst with life, no doubt thanks to the use of natural yeasts and fermentation in stainless steel and neutral acacia vessels.

More Exceptional Producers
in Trentino–Alto Adige

ALTO ADIGE

Abbazia di Novacella
Alois Lageder
Ansitz Waldgries
Baron Widmann
Cantina Terlano
Elena Walch
Franz Gojer
Köfererhof
Loacker
Manincor
Muri-Gries
Nals Margreid
Radoar
Reyter
San Pietro
Thomas Niedermayr

TRENTINO

Azienda Agricola Salvetta
Cantina Alpina
Cesconi
Fedrizzi Cipriano
Zeni

Friuli Venezia Giulia

The third of Italy's three semi-autonomous regions in the north (the other two being Valle d'Aosta and Trentino–Alto Adige), Friuli Venezia Giulia has been shuffled back and forth between kingdoms and republics from the Romans to the Venetians to the Austrians and back to the Italians so often that the borders have been all but erased. And at any rate, the modern borders, established in 1954, are of little daily import to the Friulians who live on the borderlands and blend cultures with gleeful abandon.

Friuli Venezia Giulia, like nearby Trentino–Alto Adige, consists of two distinct regions, each with its own traditions, language, inclinations, and historic loyalties. In the west is Friuli, made up of two provinces, Udine and Pordenone, and long held by the Venetians. In the east is Venezia Giulia, partitioned under the Roman emperor Augustus to include parts of what are now Italy, Croatia, and Slovenia. Slovenian is as commonly spoken as Italian here, and surnames have few of the vowels one finds elsewhere in Italy. The main city of the region is Trieste, on the shores of the Adriatic, a typically atypical crossroads city, equal parts Italian, Austrian, and Eastern European. Somewhat confusingly, because the full name Friuli Venezia Giulia is so extravagantly multisyllabic, the entire region is frequently referred to simply as Friuli. To this I will, under protest, adhere.

As a whole, the region rests upon the shores of the Gulf of Trieste, a shallow bay of the Adriatic Sea that gives at least the southern coast an almost Mediterranean climate. While most of the western provinces are occupied by a broad plain and therefore have historically been oriented more toward the production of industrial wine, the eastern provinces contain the foothills and slopes of the Julian Alps to the north and enjoy the adjacency of the Adriatic to the south. Before it receded to its current borders, the Adriatic in fact flowed over much of this land, and in some ways it still does. The soil, known as *ponca*, is the former ocean floor, a mix of sandstone and marl (a lime-rich mud), riven with marine fossils. These foothills—the Colli Orientali del Friuli in the north and the Colli Gorizano in the south—are home to the region's best winemakers and most favorable terroir.

As you might expect from a region thus tussled over, Friuli has its share of international varieties, such as Sauvignon Blanc and Merlot. But since these arrived in some cases nearly two hundred years ago, brought by the Hapsburgs, then the rulers of the Austro-Hungarian Empire, it's hard to decry them as interlopers. As you might *not* expect from a land so often traded, though, Friuli has a distinct and well-defined local winemaking tradition, especially along the southeastern border. Thanks to the noble Austrians, who crowded Trieste's streets and cafés with their courts and attendants, all demanding plates of speck and bottles of wine, high-quality winemaking has long been valued here. In fact, for years, Friuli was the diamantine heart of Italian white wine making. This took its modern form in the 1970s, when the Friulian winemaker Mario Schiopetto introduced stainless steel fermentation and began producing severe white wines whose crisp minerality could cut glass. These were soon imitated across the region, amplifying its footprint domestically and eventually globally. By 2000, these more angular wines were joined by the so-called Super Whites, blended wines from mostly French white grapes that relied heavily on the barrique for their brash international style, which came to define the 2000s. Between these two poles, there seemed little room for much else.

But that is no longer the case. If anything truly defines the winemaking landscape of Friuli, it is a relentless spirit of innovation, fearless embrace of experimentation, and a harmonious diversity of styles. Part of this is, no doubt, a consequence of turmoil. The province has always been the back door between Italy and Eastern Europe, and the hills are soaked with the blood of hard-fought battles. During World War I and World War II, much of the countryside was laid to waste. And in this void Friulians had to decide on and build their own future nearly from scratch. Though Schiopetto has stuck with his clean style, taking great pains to avoid exposing his wine to oxygen, other vineyards have continued to evolve. In the 1990s, Josko Gravner and Stanko Radikon began the opposite movement toward more natural, earthier wines. They buried themselves—and at times their wines—in the past, using modern technology sparingly to rectify hygiene issues but never to alter the ancient spirit of their vineyards. These experiments with amphorae and natural yeasts resulted in the current orange wine revolution and today, though Friuli is still home to some of Italy's best white wine, its orange may be close behind.

As for the future of Friulian wine, many see it on the Carso Plateau, south of the Colli. This difficult terrain, a few centimeters of soil over limestone bedrock, was once forest, but that was cut down by Venetians, who built their city at the expense of the area's oak. Now windswept, the land demands almost Herculean effort to cultivate it—solid rock to be broken with heavy machines, so much stone that much of it was used to build the houses and churches that sit on the plain. But men like Edi Keber saw the potential of this ungenerous surface in the late 1980s to yield hardy character-driven wines. Today, along with Keber, there's a growing group of excellent producers, including Benjamin Zidarich, Paolo Vodopivec, Sandi Škerk, and Skerlj, who embrace the labor and reap the rewards.

But whether in Carso or in Collio, all across the province, in an act of self-reclamation or simply part of Italian wine's ongoing evolution, winemakers are forswearing long-grown international varieties for natives like Friulano, Ribolla, Refosco, and Pignolo. Friuli continues in its role at the vanguard not only of Italian wine but also of the world's.

Native Grapes

FRIULANO

Friuli's reputation is built on this hardy white variety. Long called Tocai Friulano, it is now known simply as Friulano after a dispute with Hungary, which has its own sweet wine called Tokay. The stripped-down name fits regardless, for this grape, with its muscularity and crisp apple flavors, is by now synonymous with the region. Capable of both standing up to barrique and being sharpened into crisp minerality with stainless steel, the versatile Friulano is now being expressed with less manipulation, showing itself to possess vibrant mineral notes. Some small producers, like Giampaolo Venica and Mitja Sirk, are elevating Friulano with naturally grown single-vineyard bottlings, underscoring that the variety, though capable of international styles, is equally adept at expressing its own terroir.

RIBOLLA GIALLA

In the late nineteenth century, phylloxera and an insatiable market for international varieties conspired to nearly doom this ancient local variety. Despite its long association with the area—first historically documented in the thirteenth century as a wine for Venetian nobility, and later, under the Hapsburgs, as the drink of choice for the region's rulers—by the mid-1990s, Ribolla Gialla, a white grape, accounted for less than 1 percent of DOC wines in Friuli. Flaccid when coddled but capable of rising to challenges, the grape grows best in areas of poor fertility, as in Collio, its natural home. Nonetheless, while Ribolla Gialla has traditionally been thought of as producing only middling anonymous white wines, as many winemakers are now discovering, it can also yield astounding orange wine. Like Friuli itself, Ribolla is ripe for experimentation and, partially for that reason, has endeared itself to the region's more fearless winemakers. Josko Gravner became so enamored of it that he uprooted his vineyards planted with foreign grapes to focus on Ribolla and another native grape, Pignolo. The proclivity of Ribolla, known as Rebula where it grows just over the Slovenian border, to produce a textured, deep, and mineral-driven result upon skin contact is, in part, responsible for the rise of orange wines.

REFOSCO DAL PEDUNCOLO ROSSO

The Refosco family of grapes is large, with many branches and much arguing among them. But in the hills of Friuli, the most common and most prized variety is Refosco dal Peduncolo Rosso. (*Peduncolo* means stem.) After being supplanted by easy-to-grow, easy-to-sell French varieties like Merlot and Cabernet Sauvignon, the grape began its slow climb to title contention about fifteen years ago. After a steep learning curve—at first the grapes were harvested too green, and then too ripe—Friulian winemakers have found their sea legs. Today in the hands of organic winemakers like Vignai da Duline and Villa Job, Refosco is behind some of Friuli's best wines, red, white, or orange: high in tannins, high in minerality, and worth aging but too tempting to wait.

TERRANO

The jury is still out as to whether Terrano is simply a biotype of Refosco dal Peduncolo Rosso that has adapted over the years to the terrain of Istria, where it grows on the iron-rich red soils of Carso, or its own distinct variety. Called by various names (as are so many things in this linguistic Gordian knot of a region), including Cagnina, Teran, and Terran, the wines tend to be deep purple, with moderate alcohol, and characterized by high iron content and a ripping acidity. Both Benjamin Zidarich and Edi Kante make delicious and age-worthy Terrano.

SCHIOPPETTINO

The story of Schioppettino, a victim of governmental bureaucracy but the beneficiary of the efforts of enterprising locals, is a fairy tale of a long-shot comeback. Now the ancient grape, also called Ribolla Nera (though it's unrelated to Ribolla Gialla), is being expressed by scores of wineries, mostly in Udine's Friuli Colli Orientali. That it still exists is no small feat. The grape was nearly extinct when Paolo and Dina Rapuzzi heard mention of it from older Friulians. Intrigued, they began researching and found, digging through ancient manuscripts, that Schioppettino had been growing there since the Middle Ages. Unfortunately, it was vulnerable to phylloxera, which nearly wiped it out in 1855. So the couple went on a hunt, scouring the Slovenian and Italian countryside for the rare grape. Over five years, they were able to recover 60 Schioppettino vines, and within two years, they had propagated those to 3,500. However, just as the Rapuzzis were ready to claim victory, the local government excluded Schioppettino from a list of grapes allowed to be grown, all but dooming it to extinction. Outrage from the locals, followed by political organizing that culminated in legislative change, restored the grape to its rightful place in the area. When the Rapuzzis' Schioppettino did finally burst onto the market in 1977, the accolades came fast and furious, and the variety was officially rescued.

Schioppettino (or the Little Popper) does well in the hills outside Cialla and Rosazzo, where the warm Adriatic breezes, the protection of the Alps, and the ponca soil both nurture and challenge the delicate grapes to produce a gently elegant red wine with peppery, floral, and balsamic notes. My favorite iterations are those by the wineries Ronchi di Cialla and Ronco del Gnemiz (*ronco* being local dialect for the stone terrace walls carved into the hillsides). That from Gnemiz boasts the depth and complexity of a very high quality northern Rhône Syrah, but at a (much) lower price.

PIGNOLO

The Abbazia (abbey) di Rosazzo has sat atop a hill in Friuli Colli Orientali since the thirteenth century. It boasts the oldest cellar in Friuli and, in fact, held winemaking in such esteem that a fourteenth-century document threatened excommunication to those who refused to grow vines. The Benedictine monks there are the saviors of the tightly clustered, highly tannic red grape variety called Pignolo. By the mid-twentieth century, not a single commercial winery owned Pignolo vines, which were not only nearly extinct but also largely unknown. It fell to an enterprising winery worker at the abbey to safeguard the few remaining there, and he, alongside two other winemakers, succored them back to health. Now Pignolo is making up for lost time. The highly tannic red yields a full-bodied complex wine but comes on so strong it needs years in the cellar. Today the abbey's Pignolo is made by nearby Friulian pioneer Livio Felluga and by Dorigo, but it's also worth seeking out those by Ronchi di Cialla, with its local grape heroes the Rapuzzi family, who age their wine for many years before release.

The Radikon family

Winemakers to Know

JOSKO GRAVNER OF GRAVNER

The godfather of orange wine, sometimes called the Wizard of Oslavia, Josko Gravner carefully makes his way through the stone cellar of his three-hundred-year-old farmhouse on the border between Italy and Slovenia. Buried in the floor are forty-six clay amphorae, in which ferment Gravner's beloved grapes. With his closely cropped white hair, eyes crinkled behind wire-rimmed glasses, Gravner has a grave countenance but a lively, restless, and courageous spirit. Today the winery is best known for the contents of these Georgian amphorae, Ribolla Gialla, grown biodynamically, hand-harvested, and macerated for months. However, as a young man, Gravner was an early adopter of the super-clean, sharp-as-a-razor Friulian wines made famous by Mario Schiopetto. His vineyards once held mostly international varieties, and he had replaced his father's wooden casks with stainless steel vats. However, after traveling the world's wine-producing regions in the 1990s, he realized that Oslavia had its own terroir. Out went the vats, up came most of the vines, and in their stead, he planted Ribolla Gialla and Pignolo. Thus began a journey on which Gravner continues, moving ever backward to the practices of his ancestors.

SAŠA RADIKON OF RADIKON

Stanko Radikon was perhaps the first producer in Collio to make an orange wine from Ribolla Gialla in the modern age. Of course, though, as he would have told you before he died in 2016 at only sixty-two, he wasn't the first at all—he was simply following a style set forth years before. Like Josko Gravner, Radikon began his journey hewing to the orthodoxy of the stainless steel tank. But in 1995, having grown discontented with the popular but relatively featureless results, he returned to the style of his grandfather, leaving the skins of his grapes on for an extended maceration. In 1997, his first bottle of what would become known as orange wine emerged. If, in the 1990s, Stanko had prefigured the orange wine movement, in the early aughts, he anticipated the return of natural, no-sulfites wine by refusing to add the stabilizing agent sulfur dioxide to his bottles. They emerged slightly cloudy and very much alive. Sadly, Stanko passed away from cancer a few years ago. Today the vineyard is helmed by his son, Saša, who sees his role as a steward of the estate's continuing evolution. "I'm making wines the same way my father did, but not the same way he did the last time," says Saša. "My hands are different, so the wines are different." He has introduced a few innovations, including aging wines for only a year rather than his father's six, and attempting to rejoin the DOC from which his father fled in 2001—as long as, he says, they change the regulations to welcome his beloved orange wines.

EDI KEBER

Edi Keber is another of the OG Collio crews. His vineyard is in Cormons, south of the others, where his family traditionally grew Friulano. But after a moment of enlightenment in 2008, Keber and his son Kristian decided to make only one wine, called Collio Bianco, a full-bodied blend of native grapes Friulano, Malvasia Istriana, and Ribolla Gialla. Fermented and aged in cement tanks, it's Collio in a bottle. Speaking of bottles, so devoted to elevating his region's profile is Keber that he designed an instantly recognizable, lighter glass bottle with a smaller-than-normal opening that has been widely adopted throughout the region and is called simply Collio.

ENZO PONTONI OF MIANI

An air of almost mythic reverence surrounds Enzo Pontoni, whose small vineyard in Buttrio, a village in the Colli Orientali, produces some of Friuli's best wines. Tall, uninterested in publicity, and with the brooding good looks of a mid-century matinee idol, Pontoni is as elusive as a snow leopard. His ruthless cull of his own grapes is legendary, and his high standards nearly unmeetable, but his meticulous rigor has been responsible for Miani's cult status. Although he is known primarily for his Burgundian-style whites and barrique-aged Merlot, recently Pontoni has been shifting the focus of his vineyard to emphasize native grapes like Ribolla and Friulano, and he now produces a 100 percent Refosco dal Peduncolo Rosso.

SERENA PALAZZOLO AND
CHRISTIAN PATAT OF RONCO DEL GNEMIZ

Jazz vinyl lines the living room of Serena Palazzolo and Christian Patat's farmhouse atop the Rosazzo Hill in the Colli Orientali. The house belongs to Palazzolo's family, who purchased it as a weekend getaway in the 1960s. The vinyl belongs to Christian Patat, a former jazz pianist and currently a wine consultant known for his Midas touch. Today it's both home and the headquarters of Ronco del Gnemiz, the couple's much-lauded winery.

The organically grown grapes from which they make their terrifically balanced wines are both international (Chardonnay, Sauvignon Blanc, and Merlot) and, increasingly so, native. Their Ribolla Gialla is bright and age-worthy, and their three separate bottlings of Friulano—one floral, another more saline, and the third mineral—are peak expressions of the grape. Their Schioppettino is one of the best ever made.

IVAN AND PIERPAOLO RAPUZZI OF RONCHI DI CIALLA

The legacy of Paolo and Dina Rapuzzi, apostles and rescuers of Schioppettino, is now carried on by their children: Ivan, who tends to the vineyards, and Pierpaolo, the winemaker. Best known, and rightfully so, for its Schioppettino, the biodynamic winery also makes tremendous wines from seven different native varieties, including Ribolla Gialla, Refosco, Verduzzo, Pignolo, and Friulano.

BENJAMIN ZIDARICH OF ZIDARICH

No other region in Friuli confirms that character is born from hardship more than Carso, a rocky landscape where Slovenia, Croatia, and Italy meet. There the iron-rich red soil is but a veneer for dense limestone rock, and water is so scarce that during World War I an aqueduct had to be built to sustain the troops. It seems nearly inconceivable that vines can find purchase in this volcanic plateau, but the fact that some of the region's most intriguing wines are grown here remains, to me at least, a source of tremendous inspiration. This challenge has also animated men like Benjamin Zidarich, who recently carved a 12,000-square-meter cellar into this unforgiving stone. It's the least he can do to honor the hardy vines of Vitovska, a white grape that grows here, and the similarly heroic Terrano. Though he also makes a standout Malvasia, and a blend of Malvasia, Vitovska, and Sauvignon Blanc, it is Zidarich's 100 percent Vitovska that is, in my opinion, the highest expression of this unforgiving but ultimately fruitful plain.

PAOLO VODOPIVEC OF VODOPIVEC

An early disciple of Josko Gravner, square-jawed, crew-cut Paolo Vodopivec has doubled down on Vitovska as the only true expression of Carso. Like his mentor, Paolo favors Georgian amphorae, in which he ages two of his three wines, skins on. Paolo refuses to import topsoil, doesn't irrigate his vines, and plants them densely to encourage competition for resources among them. The methods may seem extreme, but when they are paired with a recent focus on precision, the results—increasingly clean and elegant wines, riven with hints of stone, salt, and minerals—more than justify the means.

More Exceptional Producers
in Friuli Venezia Giulia

CARSO

Borc Dodon
Edi Kante
Škerk
Skerlj
Villa Job

COLLIO

Borgo del Tiglio
Damijan Podversic
(Gorizia)
Dario Prinčič (Oslavia)
Franco Terpin (Gorizia)
I Clivi (Gorizia)
Livio Felluga
Mitja Sirk
Primosic (Oslavia)
Raccaro
Schiopetto
Venica & Venica

COLLI ORIENTALI

Dalia Maris
Dorigo
Le Due Terre
Meroi
Petrussa
Ronco delle Betulle
Ronco Severo
Scarpetta
Vignai da Duline

Pierpaolo and Ivan Rapuzzi

Veneto

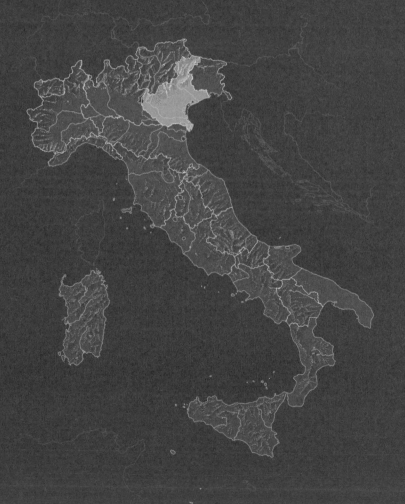

The province of Veneto has always been rather ruthless in the ways of trade. Well before it became overrun with tourists, Venice was the trading capital of the world. And one hundred kilometers to the west, Verona, best known today for being home to the world's most star-crossed lovers, was a hurly-burly marketplace in Roman times. So it makes sense that when it comes to wine, the region is famous for maximizing profit. Though deleterious in terms of quality, the move seems canny when it comes to leveraging its natural resources. The majority of the province is made up of a low-lying plain of the Piave River in the western precincts, which is perfect for industrially produced wine but a real pain for vino vero. Wines ranging from the much-maligned Soave and the similarly infamous Prosecco have driven the Veneto to become the highest-producing wine province in all of Italy. But this distinction is inversely proportional to the respect afforded it. Here is, for example, Jancis Robinson on the Veneto wine scene:

> Here lakes of pale red Valpolicella and Bardolino and watery Soave and Pinot Grigio are drained into bottles by the million for shipment to Italian and Italianate restaurants around the globe. Vineyards that are typically flat and fertile have been allowed to spew forth overgenerous yields of characterless wine with as little cachet and interest as, say, Liebfraumilch.

You don't have to know Liebfraumilch is a sweet wine from a German city called Worms to understand in what low regard the Veneto has been traditionally held.

When I opened dell'anima, in 2007, I was something of an anti-Veronisti, perhaps not as virulent as those from nearby Vicenza, who have had a century-long feud with their neighbor, but still, I had a beef. I had a hard-and-fast rule: no Prosecco and no Pinot Grigio. (I did, however, have one Soave, from Prà, and an Amarone from Corte Sant'Alda.) It's not that I hated the grapes or styles. But I thought that they had been so severely overexposed that if, as I was sure it was, my job as a sommelier was to be an agent of exploration, those well-trodden, well-known styles were supremely uninteresting.

Over the years, though, both Prosecco and, to a lesser extent, Pinot Grigio have stolen onto my wine lists. Perhaps it's that, as I've gotten older, I've grown less dogmatic. But I don't think so. The last fifteen years have seen a burgeoning movement in the outskirts of Veneto wine toward artisanal production. Being that it is ground zero for Prosecco, naturally the *col fondo* movement (see page 44) has had an outsized effect on the Veneto. More and more winemakers are awakening to the latent potential of their land. Apart from Alto Piemonte, the Veneto is home to the only volcanic soil in northern Italy. At the same time, producers such as Angiolino Maule and Alex della Vecchia and Martina Celi of Costadilà are championing their underexposed regions, native grapes, and the production of natural wine. Even in the heart of reckless Venice, natural wine is taking hold at places like Matteo Bartoli's wine bar, called, aptly enough, Vino Vero.

Native Grapes

GLERA

Until 2009, Glera was known as Prosecco, the grape that is synonymous with the style with which it is most closely associated. That the name was changed, in order to even further enlarge the permitted areas of its cultivation (while nominally protecting the OG growers), illustrates the challenges of making high-quality wine here. Everyone wants to be a Prosecco maker. (With 600 million bottles sold annually worldwide, it's easy to see why.) This has undoubtedly smudged the reputation of the variety, but in the beautiful rolling hills of Valdobbiadene and Conegliano and the hillside of Cartizze in particular, Glera is cultivated with care and attention to terroir. Producers like Silvano Follador—who is growing his Glera biodynamically, resting *sur lieviti,* and avoiding dosage to rely on the grape's natural sugars for his 95 percent Glera Brut Nature—are showing that the grape is capable of yielding an expressive, elegant, and, even at a relatively low 11.5 percent alcohol, able-bodied wine, far from the vapid fizzy water of so many imitators. Others, like Casa Belfi in San Polo di Piave, are exploring Glera as a base for both a still orange wine and a 100 percent Glera orange Prosecco.

GARGANEGA

At 16,000 acres, Soave, which is primarily planted with Garganega, is the largest white wine appellation in Italy. Naturally blessed with black volcanic basalt, which, under the right conditions, produces resilient vines and wines of deep minerality, the region has, nevertheless, become synonymous with bulk wine. Interestingly, though Soave has long suffered as the Rodney Dangerfield of the Veneto, Garganega, which first appeared in the area 1,300 years ago, has remained above the fray. This is, perhaps, because the faults of so many commercial Soaves can clearly be attributed to the reckless usage of French varieties like Chardonnay and Sauvignon Blanc or Tuscan interlopers such as Trebbiano Toscano rather than Garganega. When Garganega is allowed the space, in 100 percent or almost 100 percent bottlings, it produces a fine mineral-driven wine with a pleasing bitterness. The landscape is still dominated by corporate wineries interested more in fields of spreadsheets than of land, but long-esteemed producers like Pieropan, Gini, Inama, and Prà have never given up hope, and today they are joined by up-and-coming producers like Corte Sant'Alda, Garganuda, and Cantina Filippi, eager to restore Garganega to glory.

Garganega is enjoying an even more promising renaissance in Gambellara, just over the hills from Soave. The small region, which shares its southern neighbor's volcanic soil but not yet its commercial exploitation, is home to winemakers like Angiolino Maule, who has been a pioneer of Italy's natural wine movement since the 1990s, and Cristiana Meggiolaro, who started her natural winery in 2004. These, and other developments, make the case that Garganega, like all grapes, should be judged not by the imperfect ways it is expressed but by the nobility of its perfect expression.

DURELLA

Nearly lost in the race toward massive, oaky white wines, Durella is a rare high-acid grape with high minerality from the Lessini Mountains, the area between Vicenza and Verona. The grape's acidity, much of it derived from the volcanic soil that covers the hills, makes it especially suited for sparkling wines (which can tip dangerously into sweetness). In the sea of meh Prosecco—though, as noted, there are quite a few archipelagoes of quality these days—Durella Spumante offers a solid choice, with fresh fruit and disciplined dryness. Its recent popularity has spread the grape from the smaller Lessini Durella into wider Gambellara, where producers like Stefano Menti of Giovanni Menti, whose Durella is made without dosage, and

Grapes hung up for drying for Recioto di Soave

Cristiana Meggiolaro of Meggiolare Vini offer wonderful bottlings, the latter of which comes to its bubbles by both *refermentazione in bottiglia* and the *metodo classico*.

CORVINA

The last of the big three grapes of the Veneto is similarly known more for the styles of wine into which it is made than as a variety per se. Corvina is at the heart of the Veneto's most well-known red wines, Valpolicella and Amarone. When raised with high density and low yields, Corvina can make beautiful low-tannin, lightly colored wines with intense red cherry and sweet spice aromas. But if overcropped, as it often is, the resultant wine can be flat, and if it is overripe, the alcohol levels become overweening and unpleasant. So it demands a patient hand but then rewards the virtue.

In Valpolicella, a series of valleys on the shores of Lake Garda, running north from the Lessini Mountains southeast into the plains of the Adige River, winemaking has existed since at least the thirteenth century. In fact, the name Valpolicella means valley with many cellars. But, as is sadly the case with nearby Prosecco, the region has been diluted by an ever-expanding DOC designation. The best of these wines grows on the hillsides of the three valleys closest to Lake Garda: Fumane, Marano, and Negrar. This is salient, of course, only if enough attention is given in the vinification to allow the terroir to emerge. Happily, winemakers like Corte Sant'Alda's Marinella Camerani and Monte dei Ragni's Zeno Zignoli are doing just that, using minimal intervention and showing restraint in the field and the cantina.

Corvina also finds itself in Amarone, a style with ancient roots and also from Valpolicella. The wine is made by vinifying dried Corvina grapes to concentrate their sugar and then fermenting out their sweetness to yield a dark, heavy, thick but unusually dry red wine with a prodigious amount of alcohol. At its best, Amarone can be profound, but the style has always reminded me of a Corvette: brash, technically impressive, yet always with a slightly gauche undercurrent. Though long ascendant, the fortunes of Amarone have fallen in inverse proportion to the rise of lighter, lower-alcohol reds. However, the premier makers of Amarone, like Fiorenza Quintarelli, who took over her father Giuseppe Quintarelli's winery upon his death in 2012; Marinella Camerani, of Corte Sant'Alda; and Monte dall'Ora will continue to showcase the best the style has to offer. But if given the choice, I still prefer the rather more sprightly expression of Corvina present in Valpolicella.

If Amarone is Corvina's fugue and Valpolicella its concerto, Chiaretto is a divertimento. While much of Italy has only recently cottoned on to rosati, the Veneto has long been wise to the power of pink wine. Chiaretto, which is produced as well in neighboring Lombardy (see page 123), is here called Bardolino Chiaretto, named after the picturesque medieval town on the shores of Lake Garda. (The Veneto is also home to another still red wine called simply Bardolino, a lighter version of Valpolicella, of which Villa Calicantus makes a great example.) Whereas in Valtènesi, Lombardy, Chiaretto is made with Groppello, Marzemino, and Barbera, here it is made with Corvina (35 to 65 percent) and Rondinella (10 to 40 percent). But recently, reflective of the broader trend toward lighter reds, Chiaretto's producers have eased up. And though, in the past, I've often chosen a Chiaretto from Lombardy over one from the Veneto, makers like Villa Calicantus and Le Fraghe are making that choice increasingly difficult.

Winemakers to Know

ANGIOLINO MAULE OF LA BIANCARA

Standing in his vineyard, wire-rimmed glasses glinting in the bright light, hair cut high and tight, face crinkled into angles and creases as he contemplates rain rolling in, Angiolino Maule reminds me of Sam Shepard squinting into the sun in *The Right Stuff* or staring onto the golden fields in *Days of Heaven*. Except in this case, Maule is taking in rows of Garganega vines interspersed with clover, field beans, oats, and barley. But the intensity and spirit are the same: Here is a man on a mission.

Maule started in Biancara, where he bought a small vineyard in 1979. At the time, he and his wife, Rosamaria, owned a local pizzeria—his skills as a pizzaiolo are still renowned—but, as he tells it, he turned away from the grind of the kitchen to heed the call of the fields. (His grandparents and parents were both *contadini*, or peasants, from the Veneto.) "Wine," he says, "is the fruit that transformed earth into the culture of man." He began growing his grapes rather blindly, but in 1991, he tasted a natural wine from Josko Gravner and quickly recognized a like traveler. Since then, he's been on a mission of subtraction, trying to do as little as possible to his Garganega grapes in order to illuminate their mineral spirit.

Angiolino Maule and family

Maule quickly became part of that early crew of iconoclasts that included Stanko Radikon and Josko Gravner to the east and Edoardo Valentini to the south in Abruzzo. He recalls being first ignored and then mocked. But his biggest critic is himself. Maule is a relentless and punctilious winemaker who shares a penchant for self-flagellation with Enzo Pontoni of Miani. "I have the illness of always wanting to learn," he says. "I spend part of the year happy and part of the year knowing I could have done better."

Throughout the thirty years of La Biancara, Maule has never stopped experimenting. Inspired by Gravner and a trip to Georgia, he installed amphorae. But they didn't live up to his standards. "They can give the impression of wet earth," he tells me. So he replaced them with old Slovenian oak barrels, which allow the wine to breath but don't impart any flavor. His philosophy could be summed up as: Nothing added. Nothing removed.

What Maule realized is that without the use of chemicals, either on the field or in the cantina, or other sorts of artificial manipulation, his vines needed to stand on their own. He's dedicated the bulk of his work and the lion's share of his days to finding innovative ways to protect his precious vines without the use of pesticides or chemicals. He's minimized the use of copper, an organic treatment that others sometimes employ to prevent mold and mildew, and ceased any application of sulfur. Instead, he's experimented with orange oil, seaweed, and vegetable extracts. In the cellar, all the yeasts are natural, and his wines are not filtered.

It's hard work, clearly, but his wines justify the effort. Garganega has never shown itself as a more worthy or expressive grape. Three cru versions of Maule's Pico, a wine hailing from the highest of his vineyards, show Garganega's range: In Monte di Mezzo, it is a linear, balanced wine with ripe yellow apple fruit. In Faldeo, made from older vines, the apple flavor is joined by cinnamon and peaches. And in Taibane, from a terraced vineyard, the fruit is imposingly ripe. But the man won't stop experimenting. The latest offering is "Garg'n'Go," a pet-nat made from 100 percent Garganega.

STEFANO MENTI OF GIOVANNI MENTI

It was Easter Sunday on a hill just outside Gambellara. At the door of Giovanni Menti, the young winemaker Stefano Menti stood waving to us. In the kitchen, his wife, Katerina, had prepared an epic feast: white asparagus from Vicenza and Adriatic branzino, with a glass of "Omomorto," the very dry sparkling wine made from the Durella grape from Menti's vineyards. The scene, one feels, could have taken place centuries ago, and, one hopes, will continue to unfold for many more.

Menti is a prime example of a *ritornato*, one who strayed from the family farm only to return, in his case, out of a sense of duty and deep passion. The vineyard was founded in the nineteenth century by Stefano's great-grandfather, Giovanni. Gambellara boasts black volcanic soil, which is better for the growing of grapes than it is for the building of houses. (Menti says the stones retain so much humidity he has to repaint his house every year.) Giovanni planted Garganega and made both sweet and dry wines, as did his son Antonio, and Antonio's son, also called Giovanni. Although Stefano helped out on the farm as a boy, his path was not among the vines. In fact, he was living in the Czech Republic with his Czech wife, Katerina, and working at a water company, when he got a call from his father. The vineyard was in trouble and Gianni needed his help.

This was in 2002, when natural wines were not yet as exposed or celebrated as they are today but were beginning to catch on. Stefano decided to run the vineyard looking both forward to the future and back to the ways of his great-grandfather. He traveled around Italy visiting natural and organic vineyards, and when he returned home, he converted his own.

He eschewed pesticides and herbicides, helped, it turns out, by the fact that Garganega's thick skin makes it resistant to fungus. In 2004, the farm was certified organic. Today Stefano is an energetic and creative champion of local varieties like Garganega and Durella. He showcases the grapes in myriad ways, from Monte del Cuca, an intensely perfumed orange wine made with Garganega grapes from a single vineyard and macerated for thirty-two days with skins to a *sui lieviti* sparkler to bottles of "Vin de Granaro," a sweet wine made from air-dried Garganega.

FILIPPO FILIPPI OF CANTINA FILIPPI

The family of Filippo Filippi has lived in Castelcerino, 1,300 meters above the valley floor in Soave, since the 1300s, and they've been making wine here for more than a century. But I'm not sure if any of this ancient line made it with the free-spirited enthusiasm of Filippo. It takes a lot to devote oneself to Soave, to root deep into its murky reputation seeking treasure. But Filippo did that in 2003, when he began making his own wine by eschewing the common practices of many of his fellow winemakers. Instead of deforesting the hillsides and planting neat rows of moneymaking vines, he left the woods brambling around the vineyard. Instead of plumping out his Soave with Trebbiano di Soave, he produces 100 percent Garganega. (He also grows Trebbiano but bottles it separately.) His vines are old, dating to the 1950s, and his elevation is the highest in the region. His methods are organic and his approach is to use a light touch. Like the most esteemed natural winemakers, he sees his job as supporting nature, not taming it. This is reflected in both the vineyard, where he harvests his grapes by hand, and the cantina, where he uses only natural yeasts, allows the wine to rest on its lees, and ferments it in stainless steel tanks. One is rewarded by his wines, which display a complexity and minerality almost, and sadly, unique to Soave.

MARINELLA CAMERANI OF CORTE SANT'ALDA

Marinella Camerani wasn't born into a winemaking family, and she didn't come to her property in the Val di Mezzane by default. She was working as an accountant at her father's industrial firm when, in 1985, she decided she had had enough of P&L statements. Her father had bought an old farm a few years earlier, and Marinella moved there and began to restore it. Part of the farm was derelict vineyard. Without any formal training but with a willingness to learn, she brought life back to the farmhouse and then began planting, eventually making wine from local varieties Corvina, Rondinella, and Molinara. She'll be the first to admit her initial attempts could have used improvement. So she set about improving them.

Traveling winemaking regions from Barolo to Bordeaux, reading voraciously, consulting prodigiously, analyzing rigorously, and following her own instincts, Marinella approached the process with what in Zen is called *zazen*, beginner's mind. What she landed on is that credo of so many natural growers. As she puts it, "No excess or leaps forward . . . Only personal dedication to the search of the best possible balance between things." She shunned the use of chemicals both in and out of the cantina. She harvested by hand. She allowed the grass between the rows of vines to go wild so that the smells of clover and mint and arugula waft up to kiss the grapes.

Today Corte Sant'Alda buzzes with life. Marinella; her Peruvian husband, Cesar; and their three daughters are all involved. A cacophony of cows, pigs, and geese can be heard over the rustle of the olive trees in the wind. Corte Sant'Alda was one of the first completely biodynamic and certified organic farms in Valpolicella, and that is just one of the many ways Marinella has innovated. Her twenty-one hectares were the first under the exacting vine-training system called Guyot, and her wines are fermented in large open wooden vats using only wild yeasts. Out of her cellars come an old-style Amarone, less big-bodied perhaps than other more braggadocious reds but with infinitely more charm; bottles of stellar "Ca' Fiui" Valpolicella; and even an amphorae-aged 100 percent Molinara rosato.

MIRCO AND GLORIA GOTTARDI OF CONTRÀ SOARDA

Up the rolling hills of the northwestern reaches of the Veneto is Breganze, home to the rare white Vespaiolo. Most often its clusters of sweet white grapes—Vespaiolo is named after the *vespe*, or wasps, it attracts—are tied together and dried for a sweet wine called Torcolato. But here, on the pristine estate of the Gottardi family, they are organically grown and pressed gently to become instead a delicate dry wine with high acidity and a deceptively aromatic nose. It smells like honey and tastes like licking a stone. The winery, founded just in 2000, with its first vintage appearing in 2005, has had the benefit of using natural and organic methods from the beginning. Mirco and his wife, now joined by his children, Marcello and Eleonora, and a drove of donkeys, harvests the grapes by hand and vinifies them in what looks like a Gaudi-inspired cantina carved into the side of the hill.

ALEX DELLA VECCHIA AND MARTINA CELI OF COSTADILÀ

Costadilà, founded in 2006 in Tarzo, a small commune on the hills north of Treviso just outside the Prosecco DOC, makes some of the most interesting Prosecco in the Veneto. But to call it a winery is to undermine part of its foundational tenets. It was created by three friends including Ernesto Cattel, who sadly passed away in 2018, with the idea not only to reintroduce small-scale viticulture but also to provide a template for breaking free of the tyranny of monoculture that turns every Italian hill into a vineyard and every farm into a winery. So, though I suppose I could wax lyrical about their three *sui lieviti* sparkling wines, which come from and are named after their vineyards of variegated altitudes, and I could note how their wines, fermented in their own yeasts and free from sulfur, are made not just with the common Glera but with all four permitted grape varieties—Bianchetta, Verdiso, and Moscato Giallo, as well as Glera—I'd rather talk of the geese, goats, pigs, and cows, and of the vegetables, daisies, and wildflowers that make Costadilà so much more than a winery. Today the project is overseen by Alex della Vecchia and Martina Celi, Ernesto's colleagues and disciples, who are continuing the great work.

DANIELE DELAINI OF VILLA CALICANTUS

The tiny winery—at only eight hectares, it is one of Lake Garda's smallest—Villa Calicantus literally embodies a return to roots. Founded in 1860, Villa Calicantus was one of Bardolino's largest producers of wine for years in the mid-twentieth century. But, as the narrative so often goes, when its current owner, Daniele

Delaini, inherited the property from an aunt, he had long departed the cantina for city life. The villa was abandoned and the vines left untended. Yet after selling Italian wine in Paris, Delaini decided to start making his own. On these vineyards high on the hills of Lake Garda's east coast, he is intent on restoring Bardolino's reputation as a *vin de garde*, a wine meant to age, as opposed to the somewhat facile but immediately drinkable form preferred by Lake Garda tourists. From the beginning of this new era, Villa Calicantus has been organic, and more recently, it's become biodynamic as well. The wines—Vino Rosato "Chiar'Otto," Vino Rosso "Soracuna," and Bardolina Classico "Avresir"—are as expressive as it gets here.

ZENO ZIGNOLI OF MONTE DEI RAGNI

If you're ever lucky enough to come across some of the scarce allocation of exceedingly refined Valpolicella and Amarone from this gem of a vineyard and poet of the vine, jump at it. From five acres high upon a hill in the Valpolicella Classico region, Zeno Zignoli produces a scant 300 to 500 cases of Amarone and Valpolicella Classico a year. (The rest of his acreage is given over to cherry trees, olive trees, corn, wheat, grain, sheep, goats, and bees.) Zignoli eschews intervention and labels with equal disdain, so, though he has been farming organically and biodynamically since he took over his wife's family plot in the mid-1990s (Ragni is his wife's family's surname), you won't find any stamps of approval here. Zignoli is a self-sufficient small winemaker, tending to his fields with the help only of a trusty horse, manually crushing his grapes, then leaving them in open containers, allowing the grapes destined for his Amarone to dry in his attic, and aging his wines for well beyond the legal minimum. The fruit of one man's passionate vision, his wines are complex, deeply flavorful, and utterly unforgettable.

More Exceptional Producers in the Veneto

AMARONE/
VALPOLICELLA

Monte dall'Ora
Monte Santoccio
Quintarelli
Tommaso Bussola

BARDOLINO

Le Fraghe

COLLI EUGANEI

Ca' Orologio

GAMBELLARA

Cristiana Meggiolaro
Davide Spillare

PADUA AREA

Il Dominio di Bagnoli

PROSECCO

Ca' dei Zago
Casa Belfi
Casa Coste Piane
Col Tamaríe
Gregoletto
Le Vigne di Alice
Mongarda
Silvano Follador
Zanotto

SOAVE

Anselmi
Coffele
Garganuda
Gini
Inama
La Cappuccina
Pieropan
Prà

Emilia-Romagna

If the wine scene in the Veneto is dominated by Prosecco and Soave, that of neighboring Emilia-Romagna suffers a similar fate under the powerful scarlet shadow of Lambrusco. And yet Emilia-Romagna, as is true of the Veneto, has much to offer the keen-eyed connoisseur. Even more important, it has treasure to offer the enterprising independent-minded winemaker too. While the eyes of the wine world continue to roll at the millions of gallons of industrial Lambrusco flowing from the province, Emilia-Romagna has quietly become an artisan-wine powerhouse.

The region, sixth in size, extends diagonally to the northwest from the Adriatic coast near Rimini past the hills of Piacenza, to almost touch the Mediterranean in the west. (Only a sliver of Liguria frustrates that peninsula-spanning ambition.) With nearly 150,000 acres under vine, it is one of Italy's most productive regions. About half of the vineyards occupy the Ferrara plain, with the other half in the foothills of the Apennines to the east. Across nearly the entire province flows the stately Po River, starting in the Cottian Alps and ending at the Adriatic Sea.

Like other hyphenated provinces (or, in the case of Friuli Venezia Giulia, formerly hyphenated provinces), Emilia-Romagna can be neatly cut into two culturally and geographically distinct regions, lassoed together not by the laws of nature, but by those of man. Emilia—which consists of the western provinces Piacenza, Parma, Reggio-Emilia, Modena, Ferrara, and Bologna, west of the Savena River—belongs firmly to northern Italy and its mountain peoples; Romagna, consisting of the coastal provinces Rimini, Forli-Cesena, and Ravenna, has more in common with central Italy. This bifurcation is clear in both language (the courtly Emilian dialect versus the vowel-cutting Romagnol one) and cuisine (delicious versus extremely delicious). Many of Italy's most famous foods hail from this region, including prosciutto di Parma, Parmigiano-Reggiano, ragù Bolognese, tortellini in brodo, aceto balsamico di Modena, not to mention Mantuan melons, Po River delta rice, asparagus from Altedo, culatello di Zabello, mortadella di Bologna, and the nectarines of Ravenna.

Much of this was traditionally washed down with Lambrusco, a style of dry sparkling red wine made with several clones of the red Lambrusco grape, the most important being Grasparossa di Castelvetro and Salamino. The style has a long history in Emilia-Romagna, dating back to the Romans, and was name-checked by Virgil, from nearby Mantua, as well as Cato and Pliny the Elder. Traditionally Lambrusco grapes were grown in the gentle plains between Modena and Reggio-Emilia. Today, though the province of Modena is still the epicenter of Lambrusco production, it often seems that wherever the land is not occupied by wheat or corn, Lambrusco vines grow. The wine's modern dominance began in the 1970s, and as its international popularity grew, it became increasingly sweet and disconcertingly anonymous. Today there are about 170 million bottles of Lambrusco produced a year, making it, by far, the region's dominant wine. Many of those bottles come from companies like Riunite, one of Italy's largest wine producers. The company's facilities just outside Parma occupy a plant the size of a sprawling airplane hangar. And though I've never been inside, I imagine some sort of sterilized scene out of the *Mister Rogers' Neighborhood* episode "How People Make Crayons" (episode #1481), but less colorful and less fun. It's then shipped around the globe, and as it travels, it unfortunately spreads the message through its bottles that Lambrusco is nothing more than a sickly sweet trifle.

Nothing could be further from the truth. Like all wines, Lambrusco is capable of expressing its land if given the opportunity to do so. Increasingly, it's being given that opportunity by artisanal producers such as Cinque Campi and Fattoria Moretto. The first step in the style's rehabilitation was to claw back from industrial producers the methods of secondary fermentation. Logically on such a large scale, the majority of producers used the Martinotti method (see page 42), but in the 1990s, a small group of outliers returned to the more character-driven *metodo classico*. However, the process is expensive and thus out of reach of many smaller winemakers who, though passionate, lacked the resources to pursue it.

Much more accessible is the *pétillant naturel* (pet-nat) method (see page 44), which has the added benefit of encouraging the nature of the grapes to have its say. In Emilia-Romagna, this method is called *rifermentazione in bottiglia,* and though there are slight differences in whether and when fermentation stops and restarts, the effect is the same. The return to the ancestral methods, which are not optimized for bulk production, soon began to yield complex, intriguing examples. Bottles from Lambrusco producers such as Graziano and Luciano Saetti are among my favorite of all spumanti in Italy.

But Lambrusco is not the only sparkling wine Emilia-Romagna has to offer. Perhaps, with the wine world either distracted by Lambrusco or dismissive of it, the region's other winemakers have felt freer to experiment in the margins. Producers such as Ancarani, Croci, Cà de Noci, and Mirco Mariotti, to name a few, are producing a truly astonishing array of pet-nats from lesser-known local grapes like Ancellotta, Malbo Gentile, Albana, and Malvasia di Candia. These wines—which come in a sunset's array of hues from deep red to vibrant orange to light pink and near-translucent white—are a potent reminder to take note of the often neglected. And although I wouldn't put money on pet-nats supplanting industrial Lambrusco anytime soon, the rate of change is increasing exponentially. As Elena Pantaleoni, the legendary natural wine pioneer at La Stoppa, pointed out to me, now that there are more people to bounce ideas off, there's an opportunity for the region to transform more quickly.

As one might well expect, with a terrain as varied as that of Emilia-Romagna, not all of the land has been devoted to the cultivation of Lambrusco varieties. At the northwest border are the Colli Piacentini, the well-aerated hills outside Piacenza near Lombardy, of which says winemaker Massimiliano Croci, "There's no Lambrusco here." Croci grows red grapes such as Barbera and Croatina and local white grapes including Malvasia di Candia Aromatica, Moscato Bianco, and an antique variety called Ortrugo. In these hills too, you'll find Elena Pantaleoni's La Stoppa, which has become a benchmark for natural wine with their deeply earthy "Macchiona," made with half Barbera and half Bonarda, and their fresh, funky "Trebbiolo," made with 70 percent Barbera and 30 percent Bonarda. The land is also ideal for earthy Barbera, and Danila Mongardi of Al di là del Fiume has gone the Gravner route, making a unique, earthy, biodynamic 100 percent Barbera, aged in amphorae, called "Dagamò."

Even in often-overlooked Romagna, a province with its eyes turned toward the tourist-lined beaches, winemakers are pushing the region forward. Claudio Ancarani of Ancarani wines, in the foothills of the Apennines, is just one of eight producers focusing on the nearly extinct Centesimino grape. Andrea Bragagni, in the medieval village of Brisighella, hand-harvests Albana and Famoso, and Chiara Condello is the ascendant superstar of Sangiovese.

Whether they are the beneficiaries of salutary neglect or simply capitalizing on an untapped market, the story of Emilia-Romagna's vino vero winemakers is, to me, one of the most inspiring, for it demonstrates that in a region dominated by industrial producers, small human-scale makers, those motivated by passion and the call of the land, can still triumph. Small victories to be sure, by the bottle, not the case, but victories nonetheless.

Native Grapes

LAMBRUSCO

Lambrusco isn't a grape, it's a family of grapes, each with subtly but substantially different characteristics. As a blended wine, Lambrusco is a carefully cast mix of these varieties, which include Lambrusco Grasparossa, Lambrusco Salamino, Lambrusco di Sorbara, and more. Among the big three are Grasparossa, which prefers hillsides to plains, is thick-skinned, and yields dark structured wines; Salamino, which takes its name from the salami-like shape of its berries, acts as the pollinator for Lambrusco di Sorbara, and yields lighter sprightly notes; and Lambrusco di Sorbara, one of the oldest varieties, which prefers sandy soils, is widely grown, and is so aromatic that it is sometimes called Lambrusco della Viola, or of the violet. Other notable family members include Maestri, which, when not overgrown, yields the fruitiest, creamiest of all Lambruschi, and Fiorano, a rare variety that grows near the town of same name outside Modena.

PIGNOLETTO

Known by many names, Pignoletto is also called Grechetto di Todi, Rebula, Rebolla, Grechetto, Pignulet, and more. Using the name of the recently created DOCG, the white Pignoletto grows in the hills just outside of Bologna, the Colli Bolognesi. Although the sparsely grown grape yields a light, minerally still wine, its high acidity makes it a promising candidate for both frizzante and, more rarely, sweet wine. One of the more versatile of the region's white wines, Pignoletto is expressed as a pet-nat in the hands of Federico Orsi of Orsi Vigneto San Vito.

ALBANA DI ROMAGNA

It takes a certain chutzpah to devote oneself to Albana di Romagna, a grape that has become synonymous with the perfidy of the DOC system. When the denomination was created in 1987—Albana was the first white wine in Italy to achieve DOCG status—many perhaps rightly cited it as an example of the debasement and politically motivated origins of the wine pyramid. Could it really be the case that the little-known and middling Albana di Romagna deserved to be considered equal to Barolo and Brunello di Montalcino? Though it is true that Albana, which flourishes in the same corner of Romagna as the mighty Sangiovese, was not at the time perhaps deserving of such elevation, over the intervening years, winemakers like Claudio Ancarani and Andrea Bragagni have coaxed wines of deep character and distinct flavors from these grapes. Especially with those grapes grown on the hillsides of the Apennines and cultivated by dedicated winemakers, Albana in fact shows itself deserving of our attention and esteem.

FAMOSO

In a gleeful bit of grape-geek irony, the grape known as Famoso was thought to be extinct until the year 2000, when two rows of it were found in the province of Forlì-Cesena. Although it had been famous in the 1400s, by the 1900s, the white grape, with its aromatic florality, had run afoul of the times, its decline further hastened by phylloxera. A hardy, productive variety, Famoso is enchantingly sensitive to its terroir: a wine made with Famoso from Cesena and one with grapes from Faenza will bear marked differences. Today the grape is cultivated by a small cadre of vino vero winemakers, including native-varietal champion Claudio Ancarani, whose Famoso wines retain acidity, and Villa Venti, an idyllic vineyard between the villages of Roncofreddo and Longiano, where it makes its way into "Serenaro," bright, crisp, and fresh.

SANGIOVESE DI ROMAGNA

A variant of its better-known Tuscan cousin, Sangiovese di Romagna grows south of Bologna, heading along Via Emilia to the hills and cities of Faenza, Forlì-Cesena, and Rimini and then toward the Apennine foothills and Adriatic Sea. Though capable of the bold expressions of, say, Chianti Classico, Sangiovese di Romagna often achieves ripeness without overbearing tannins, yielding a fleet-footed red wine much deserving of its recent increase in popularity. Although the grape has been grown here since the seventeenth century, the renaissance of Sangiovese di Romagna is still embryonic. Most of the artisanal winemakers devoted to its cultivation started their journeys in this century. The DOC has existed since 1968, but production had been marred by an overreliance on the co-op system. Yet Sangiovese di Romagna has proven itself tremendously expressive, as evidenced in wines from Andrea Bragagni's raw, savage, and earthy bottles to Chiara Condello's bright, crunchy, and fruity ones. The best producers have banded together to form a *convito* (consortium) to promote, protect, and cultivate this up-and-coming grape.

Winemakers to Know

LUCIANO AND SARA SAETTI OF VIGNETO SAETTI

Luciano Saetti and his daughter, Sara, are among the small minority of winemakers in Lambrusco fully committed to expressing both terroir and the single varieties of the grape so often lost by blending. The family's cantina near Carpi, a commune north of Modena, produces some of the liveliest wines of the region. A small operation, Saetti uses the low-cost, high-reward *rifermentazione in bottiglia* technique. Everything involving these wines—from harvesting the vines to (spoiler alert) attaching a secret message underneath the lid to affixing a fabric label made by local artisans—is done by hand.

Unlike most larger Lambrusco houses, Saetti focuses on showcasing single varieties. In the case of Il Cadetto, for example, the Saetti use 100 percent Lambrusco Salamino di Santa Croce from vines dating to 1964 is used. The grape, most often blended with the much more common Lambrusco Grasparossa, shines here. "Some people dismiss this grape, claiming it lacks aroma. It's austere, but it also smells of berries and has a pleasant bitter finish that comes directly from the grape skins," says Sara. "It expresses itself in this wonderful way if you let it."

VITTORIO GRAZIANO

Vittorio Graziano seems ripped from the lyrics of a Bruce Springsteen song, that is, if the Boss considered the hills outside of Modena his métier. The son of a furniture maker, Graziano was well on his way to a Walter Mitty existence, toiling in cubicle-obscurity as an administrative accountant. In the 1970s, his twenties were being sucked dry by thoughts of his fifties. Then his mind turned to the half-hectare of farmland his father owned in the village of Castelvetro di Modena. What he really wanted to do, he realized, was to walk in the sun, so he quit his job, knowing he was born to run . . . a small vineyard of native grapes.

Graziano didn't go to a fancy oenology school. Instead, he asked the *contadini* what grapes grew best there and then experimented with them. A shirker of labels, eye-poker of authority, he has always rejected the primacy of the wine pyramid, making his Lambrusco under the catch all IGT label. IGT or DOC, the acronyms don't matter to him. One taste of his pet-nat—deeply savory, structured, and a little wild—redefines what Lambrusco can be. His now ten hectares of vineyards are organic, though he won't bend his knee to certify it, with grass growing like crazy, legumes in profusion, and varieties including Grasparossa, Malbo Gentile, and a bunch whose origins even he doesn't know. There is no better argument for throwing off the shackles of both the office world and external validation than the life and wines of Vittorio Graziano.

GIOVANNI AND ALBERTO MASINI OF CÀ DE NOCI

To the extent that Spergola, a high-acid white grape, was thought of at all, it was mistaken for a lesser variant of Sauvignon Blanc. However, Vittorio Masini, a doctor of agriculture, didn't agree when he planted his small walnut farm (Cà de Noci means House of Walnuts) on the hillside of the Crostolo Stream valley in 1970 with Spergola, along with Malbo Gentile, Monterrico, and other rare native grapes. Since 1993, Vittorio's sons, Giovanni and Alberto, have continued their father's work, always organically, always by hand, vinifying Spergola using both the *metodo classico* and as a pet-nat, as well as producing a rare and excellent Lambrusco Sottobosco from Malbo Gentile and Monterrico. Lambruschi of such harmonious balance between acidity and tannin are found but seldom. Cà de Noci's Notte di Luna—a blend of Moscato, Malvasia Candia di Aromatica, and Spergola fermented together on the skins for ten days—is fragrant with elderflower and honey aromas, matched by outstanding acidity from the Spergola.

ELENA PANTALEONI OF LA STOPPA

From the hills outside Piacenza in the far west of Emilia-Romagna, Elena Pantaleoni is one of the most powerfully charismatic voices of the natural wine movement. As a boy, her father, a printer from Piacenza, passed the old farmhouse that had once belonged to a winemaking lawyer but had fallen into disrepair and fell in love with it. In 1973, he bought the land and built his own cellar. But he left the vines—mostly French varieties planted by the Genovese lawyer from whom he purchased the property. Meanwhile, his daughter, Elena, built a life for herself in the city as the owner of a bookstore and record shop. When her father died in 1991, Elena inherited the vineyard, its land, and—importantly—the winemaker, Giulio Armani, who had been in the employ of the elder Pantaleoni since 1980. Guided by Elena, the two charted a new course for the vineyard, grubbing up the French vines in favor of native varieties of Barbera, Croatina (called Bonarda here), and Malvasia. "It took me some years to understand what I wanted to do," she says. "I finally decided that I wanted to make wines with more identity. It's important to understand where you are and to make wines that reflect where you are." La Stoppa turned hard toward nature, in both the vineyard and the cantina. Whereas many winemakers shy away from volatile acidity or naturally occurring

Brettanomyces (yeasts), Pantaleoni leans into them, professing that what many consider defects "let you make more drinkable and more elegant wines."

Today Pantaleoni is focused on polycultural farming, much like her contemporary and friend Elisabetta Foradori. (In fact, Foradori's daughter, Myrtha, is consulting on Pantaleoni's enormous garden bursting with tomatoes, borage, and cucumbers.) And despite her inspired advocacy for natural wines, it is really her bottlings—including a vivid, buoyant 70 percent Barbera and 30 percent Bonarda blend called "Trebbiolo" and another, "Macchiona," a 50/50 blend of the two with strong, silky tannins, named after the farmhouse that sits amid her orchards—that continue to be her most convincing argument. Her orange wine, Ageno, is named after the former owner, Giancarlo Ageno. Though I wonder how he'd take the thought of his precious Chardonnay being ripped up for Malvasia Candia di Aromatica, which stays on its skins for five months, I'm sure he—like the rest of us—would flip for the wildflowers on the nose and the citrus and apples on the palate.

GIULIO ARMANI OF DENAVOLO

Denavolo is winemaker Giulio Armani's mountain workshop. When not in the cantina or vineyards of La Stoppa, where he's Elena Pantaleoni's partner-in-arms, he's here, on the mountainous soil of the Colli Piacentini, relentlessly experimenting with growing rare local varieties of white grapes like Ortrugo and Malvasia di Candia, from which he makes three skin-macerated wines: "Catavela," the lightest and freshest; "Dinavolo," the deepest and most macerated; and "Dinavolino," which rests between the two. All three benefit from the high elevation, which allows the grapes to ripen yet yield wines with a restrained 11 percent alcohol. All three, as one could expect of a wine by Armani, are concise, joyous expressions of this small bit of hillside on the plain of the Po River.

FEDERICO ORSI OF ORSI VIGNETO SAN VITO

Born in Bologna but raised in Brazil, Federico Orsi was well on his way to a life as an international businessman when he began studying for his MBA at the University of Bologna. But in 2005, he saw a farmhouse in the hills in the town of Valsamoggia in the Colli Bolognese and fell in love. Nearly twenty years later, that farmhouse has become the winery, farm, agriturismo, and burgeoning mortadella business Orsi Vigneto San Vito.

Federico Orsi

The twenty-five-acre property sits on a hill, with its vineyards unfurling from the cantina at the top. When Orsi bought the land, still a farm, it was planted largely with international varieties like Sauvignon Blanc and Riesling, with some Pignoletto. Over the years, Orsi has been gradually replacing the international varieties with local ones, including Negretto, Alionza, Albana, and Malvasia (the cuttings for which he got from La Stoppa) for whites and Barbera and Negretto for reds. With nearly zero experience making wine, he found inspiration not only in the work of Radikon, Gravner, and La Stoppa, but also, as he says, from the documentary *Mondovino*, which caused him to question what he'd spent his life drinking. He resolved to express his little corner of Bologna through vines.

Today the farm is both biodynamic and organic. In the cantina, natural yeasts are used and nothing is filtered. Orsi ferments his Pignoletto into a nicely mineral *sui lieviti,* as well as a still wine from the oldest vines on the property. His Barbera, aged in neutral *botti,* shows itself. But for me his most intriguing wines are the Posca line. Also called solera wines or *vino perpetuo*—perpetual wine, which has a lovely poetic ring to it—these are aged vintage upon vintage, variety upon variety, and added to a large tank. The bottles are drawn from about 5 to 10 percent of

this. Orsi makes two: a Posca Rossa and a Posca Bianca. The Rossa contains every vintage since 2008; the Bianca, every one since 2010. This ever-changing wine is an accumulation of terroir through time, a wild exploration of what the hills of Valsamoggia have to offer. Like a mother dough, the *vino perpetuo* offers both a continuous link to the history of Vigneto San Vito and a bridge to its future.

MASSIMILIANO CROCI OF CROCI

Standing on the hillsides of the Colli Piacentini, young Massimiliano Croci is surrounded by history. The land was purchased by his grandfather and run as a dairy farm until 1970, when industrial agriculture rendered dairymen like the Crocis obsolete. So Massimiliano's father planted vines and began producing sparkling wines, availing himself of as much modern technology as he could. Although chemical pesticides were too expensive, yields were high, and the fermentation was industrial. Now the winery is under the protection of the younger generation and Croci is returning, nigh on ninety years later, to the traditions of the lo-fi approach of his grandfather.

On the other hand, the destiny of Croci's parcels might well have been determined five million years ago, as the Adriatic washed over these lands, depositing sand and seashells that now foster hardworking vines that form the soil for Croci's Ortrugo, Malvasia di Candia, Barbera, and Bonarda. He hand-harvests the vines and ferments using the pet-nat method, which follows the natural cycle of the seasons. (The winter arrests the fermentation; the spring restarts it.) The results combine the technological advances over three generations—especially when it comes to hygiene—with the spirit of his grandfather and ancestors before him.

CLAUDIO ANCARANI OF ANCARANI

In every region of Italy, in every corner, in every variance of hills, there should be a winemaker like Claudio Ancarani, someone for whom rescuing native grapes from obscurity or extinction and cultivating them takes on an almost maniacal edge. In eastern Emilia-Romagna, near Faenza, Ancarani has made it his life's work to extol both maligned local grapes like Albana di Romagna and Sangiovese di Romagna and the extremely rare Famoso and Centesimino. Centesimino was thought to be entirely extinct, until a few vines of it were found in the 1960s in the walled garden of a horticulturalist. Ancarani's Centisimino is an aromatic floral red with dark fruit, lifted by a saline minerality. His Famoso combines delicate floral aromas with minerality on the palate and a crisp finish. And both his Albana and Sangiovese di Romagna are templates for honest and straightforward, but nevertheless complex expressions of terroir.

More Exceptional Producers
in Emilia-Romagna

EMILIA

Al di là del Fiume

LAMBRUSCO

Camillo Donati

Cinque Campi

Claudio Plessi

Cleto Chiarli

Fattoria Moretto

La Collina

Lini 910

Podere Giardino

Terrevive

NON-LAMBRUSCO PET-NAT

Crocizia

Mirco Mariotti

Montesissa Emilio

Podere Pradarolo

ROMAGNA

Andrea Bragagni

Chiara Condello

Villa Venti

Le Marche

Marche, The Marches

For some reason, the addition of the definite article before the name of a place lends it a certain poetic resonance. Would The Meadowlands, home of the New York Giants, be so poetic if it were simply Meadowlands? Would Le Marche sound so epic if it were simply Marche? In the former instance, certainly not. In the latter, probably. The name Le Marche comes from the medieval word *marça*, meaning mark or march—in this case, a demarcation between two kingdoms. In other words, a borderland. True to its name, this long stretch of wild coastline along the Adriatic today marks the border between the sea and the interior provinces of Umbria. But during the period when it got its name, in the tenth century, the area represented the much-contested edge of the Papal States, with Abruzzo to the south.

Thanks in part to the strategic importance of the port of Ancona, its capital and only large city, Le Marche has been fought over and occupied from the time of the Greeks to the Romans—Julius Caesar hastened here after crossing the Rubicon in nearby Romagna—to the Lombards, popes, Austro-Hungarians, and, finally, in 1860, the Italians. Through the elbow-shaped port surrounded by limestone cliffs have flowed immense amounts of trade: from the purple dye of ancient Syracusans to today's hordes of passengers disembarking from engorged cruise ships destined for, or arriving from, the eastern shores of the Adriatic.

There is little flatland in Le Marche, which is five times longer than it is wide. From the cliffs surrounding Ancona, the terrain makes its way somewhat steeply upward to the Apennines and the border between Le Marche and Umbria. There is, of course, hinterland, but even that is within a stone's throw of the Adriatic, and, at any rate, a series of rivers starting in the Apennines and flowing easterly serve as ties to bind the province irrevocably to the sea.

For the most part, all eyes in Le Marche turn toward the water. At the table, Anconans feast on langoustines, mantis shrimp, scallops, and a virtual aquarium stew called *brodetto*. And, for at least the last five hundred years, on their tables have also sat bottles of wine made with one of Italy's finest white grapes, Verdicchio. The story of Le Marche is thus the story of this grape, its rise and fall and rise again. In a narrative parallel to those of Lambrusco in Emilia-Romagna and Soave in the Veneto, in the modern age, the once-noble variety suffered years of industrial desecration. Starting in the 1970s, Verdicchio was made widely, maudlin and kitschy, sold in the fish-shaped bottles I remember from the red-sauce

joints in Queens. That Verdicchio continued to be at least somewhat well regarded, despite its debasement, is testament to the grape's inherent strength.

In many ways, Verdicchio is the ideal grape. It boasts high levels of minerality and acidity, especially of the tartaric acid that almost inevitably ensures its longevity. And, most crucially, Verdicchio shares a quality with Riesling and Chardonnay: an artist's sensitivity to its place and the capacity to express it. A Verdicchio grown on the slopes of one side of the Matelica valley will taste very different from a Verdicchio grown on the opposite hillside. But over the last twenty years, as production has tilted back toward the artisanal, Verdicchio has continued to impress me with its changing lights and expressivity. Today, clustered around the two most well-known Verdicchio zones, Castelli di Jesi and Matelica, both in the limestone-and-clay hills of the north, the grape is enjoying a full restoration.

But Le Marche isn't only Verdicchio. The variety has recently been joined by two unique white grapes in the (somewhat meaningless but nevertheless useful) pantheon of DOCG status: Pecorino and Passerina. Both of these vervy whites hail from the commune of Offida, a small town between two rivers in the outskirts of Ascoli Piceno. In terms of reds, Le Marche offers Sangiovese and Montepulciano, both of which flourish on the southern hills toward Abruzzo and are often intergrown and intermixed. Other grapes worth mentioning include Lacrima, though it's a bit too cloying for my palate, and the savory Vernaccia Nera, a rare red grape capable of making wonderful sparkling, dry, and sweet still wines.

Native Grapes

VERDICCHIO

Recent research suggests that Verdicchio originated not in Le Marche but in the nearby Veneto, where it is today known as Trebbiano di Soave. According to that theory, the grape arrived along with horses, seeds, and hopes in the fourteenth or fifteenth century after the Black Plague had decimated the hills of Le Marche, leaving a vacuum to be filled by Venetian pioneers. No matter its exact point of origin, Verdicchio flourished in its new home. And that speaks to the grape's adaptability.

Like a well-mannered houseguest, Verdicchio accommodates itself to its environs. But of course not all house rules are equally pleasant, nor are all hillsides equal. Verdicchio thrives in the northwest precinct of Le Marche, among the terrifically jagged peaks and valleys of the Apennines. Here two appellations—one bigger, one smaller—play host to the vines. Unusually for a sea-bordering province, the vineyards of the smaller Matelica are farther up the hills of the Apennines and produce lighter, brighter wines. In this valley, the vines are planted at between 1,000 and 1,500 feet above sea level. Nights are cold, days are hot, and the difference betwixt the two—the diurnal shift—yields crisp mountain whites whose complexity develops slowly but surely before their October harvest. Verdicchio from Matelica tend to be light and lifted. If not sharp, it is at least pointed, with citrus notes and bright acidity. Many of the wines are drunk young, but that hard-earned acidity also makes them well suited to aging. More and more winemakers, such as Collestefano, are allowing their Verdicchio to rest on its lees during fermentation, adding further complexity to the wine.

Thirty miles closer to the Adriatic is the Castelli di Jesi, which produces ten times the amount of wine as Matelica. Less austere a grape than its inland cousin, Verdicchio di Castelli di Jesi is raised under the soothing influence of the ocean breezes and at a lower elevation. Grown if not in abundance, then at least in more comfort, these grapes tend be fleshier affairs, softer, rotund, and bursting with fruit, but they nevertheless retain Verdicchio's signature acidity and minerality.

As mentioned earlier, the travails of the variety are man-made in origin. It's rarely an auspicious sign that a book is best known by its cover or a wine by its bottle. In 1953, at the beginning of the *boom economico*, a Verdicchio producer from Castelli di Jesi named Fazi Battaglia commissioned a Milanese architect, Antonio Maiocchi, to create a unique bottle to promote his wine. Inspired by both the amphorae in which the ancient Etruscans transported Verdicchio throughout their empire and the silhouette of actress Gina Lollobrigida, Maiocchi presented Battaglia with a sinuous bottle he called Titulus. It didn't really look like an amphora or a sex symbol, but it also did not look anything like other bottles of wine. Its success was immediate, but the popularity of the form was double-edged. As Verdicchio production skyrocketed, its quality and reputation suffered. That debasement lasted well into the early 2000s. Yet the remedy might have been worse than the illness. To promote higher-quality cultivation, the European Union incentivized growers to grub up their vines. But, according to Riccardo Baldi of La Staffa, many chose to uproot their old vines, preferring instead the youthful productive young ones. Today, he says, the oldest vines he can find are from 1972, and many are much younger than that.

However, as one might surmise from the vast number of Verdicchio producers now rejecting the outlandish bottles, the focus of Verdicchio has returned to the wine, showcasing its ability to age much better than those bottles did. Biodynamic makers like La Staffa age their Verdicchio Riserva Selva di Sotto several years before releasing their deeply complex wines. San Lorenzo's "Il San Lorenzo" bottling is rested on lees for 110 months, then aged in concrete and stainless steel for more than ten years. I've had 1990s Verdicchios from the iconic producer Ampelio Bucci that were alive and singing, and others from La Distesa that are beautifully textured. No longer simply a light, fresh young white wine, Verdicchio is inspiring—and rewarding—producers devoted to illuminating its multifaceted charms.

PECORINO

If Verdicchio arrived from the Veneto but found its true home in Le Marche, Pecorino's journey was the inverse. The grape likely originated in Le Marche, but it is better known today in Abruzzo. The small-fruited indigenous variety has been grown in the mountains of Le Marche since the tenth century, but it was always under the radar. It was not until the 1980s that a winemaker named Guido Cocci Grifoni, of Grifoni Vineyards, planted a few cuttings gleaned from an Abruzzian farmer in his vineyard near Offida. Not quite as crisp nor acidic as Verdicchio but nonetheless worth attention, Pecorino yields medium-bodied wines with restrained fruitiness, and the grape has been enjoying a much-deserved renaissance in the last few years.

PASSERINA

Passerina seems like a variety so overlooked it may now resent the intrusion. Part of the difficulty of examination is that the term seems to refer to a range of grapes with little in common with each other, grown in Le Marche, Abruzzo, and Lazio. The version from Le Marche presents itself as a more gentle companion to Verdicchio's virtuosity. With light herbal and floral notes, the best iterations of the grape are made by Pantaleone, just outside Ascoli Piceno, and Tenuta Cocci Grifoni, on the banks of the Tronto River.

MONTEPULCIANO AND SANGIOVESE

Though the reputation of Le Marche rests on its white wines, its production of red wine nearly equals that of Verdicchio. Montepulciano and Sangiovese, either in tandem or individually, represent about 28 percent of the total production. The two varieties grow upon the Conero Massif, an imposing promontory just south of Ancona. As one approaches Abruzzo to the south, Montepulciano predominates.

Wines from the Rosso Conero DOC, which are among the best known, allow for 85 percent Montepulciano and, in general, are fuller-bodied than their Abruzzian counterparts. Though I love the work of winemakers like Alessandro Starrabba at the five-hundred-year-old winery Villa Malacari, who capture the tenderness of Montepulciano, at times Conero wines seem as massive as the rock from which they come. I prefer the reds from Rosso Piceno, a subregion south of Conero, where the higher percentage of Sangiovese yields lighter-bodied, delicate wines.

VERNACCIA NERA

Just south of Castelli di Matelica and its renowned Verdicchio are a few vineyards of an obscure, once nearly extinct, and now mildly resurgent red grape called Vernaccia Nera. Like most grapes in the Vernaccia family, it has little to do with the others. Nor is it, as is sometimes claimed, identical to Grenache. Instead, centered around the village of Serrapetrona in the foothills of the Apennines, Vernaccia Nera production proclaims its own individuality with spicy, earthy wines with restrained tannins. Although only ten hectares were included in the DOCG when it was named in 2017, its cultivation has slowly increased. The grape is traditionally made into a sparkling wine (a sort of backwoods Lambrusco), but wonderful still and dessert wines are produced by winemakers such as Colli di Serrapetrona and Alberto Quacquarini.

LACRIMA

Intensely fragrant and little known, the Lacrima grape grows in a small pocket of towns west of Ancona called the Morro d'Alba. In today's exhaustingly chronicled vinous landscape, the wines into which it is made—a fruity, almost Beaujolaisian dry red and a Passito-style sweet one—suffer from the numerous homonyms for the grape to which this one bears no relation and from the much more well-known Alba in neighboring Piedmont. The grape's name, meaning "tear," comes from either the grape's teardrop shape or the thin-skinned fruit's tendency to tear and "cry" its juice. Once common in medieval times, Lacrima eventually became so rare that by 1985, there were only 18 acres under vine. Now the number is close to 750, with numerous producers devoting themselves to its virtues. Even more hearteningly, while Lacrima was once overpowered by Montepulciano or, less frequently, by Verdicchio in blended wines, today it's standing on its own in 100 percent single-varietal bottlings, notably those by Conti di Buscareto and Stefano Mancinelli, and in a lovely rosé by Lucchetti.

Winemakers to Know

RICCARDO BALDI OF LA STAFFA

One of Le Marche's winemakers most dedicated to proving the age-worthiness of Verdicchio is also, somewhat ironically, one of its youngest. Riccardo Baldi, who grew up in Castelli di Jesi, started his winery, La Staffa, at only nineteen years old. Slowly expanding his acreage from five to twelve over the course of five years, he's also migrated toward natural methods. Today his farm is organic and his Verdicchio fermented using only indigenous yeasts. He ages and ferments his wine in old concrete tanks. Baldi is boldly stretching Verdicchio's character. His pet-nat, a mixture of Verdicchio and Trebbiano Toscano, is stony, with notes of green apple, and called "Mai Sentito!," which means, "I've never heard of that," a phrase Baldi was often greeted with when describing his vision. And, perhaps because he has so much time ahead of him, Baldi has also begun aging his Verdicchios, including an impressively full-bodied Verdicchio Superiore Rincrocca, which is aged for three years in concrete.

Riccardo Baldi

CORRADO DOTTORI OF LA DISTESA

A self-described psychedelic winemaker, the goateed, wild-maned Corrado Dottori has been called the Frank Zappa of orange wine. In the Castelli di Jesi, on the hilltop Cupramontana, once known as the "capital of Verdicchio," he fearlessly experiments with rare fifty-year-old vines of Verdicchio as well as Trebbiano, Sangiovese, Montepulciano, and Pecorino on his fully biodynamic farm. Dottori pushes the always-accommodating Verdicchio to new territory on his scant acreage, especially with his skin-on blend of Verdicchio, Trebbiano, and Malvasia called Nur. But he also turns out a few bottles of a rosé made with the maceration of both red and white grapes; a solera (or *vino perpetuo*) wine; a sparkling red; and a lively Trebbiano and Verdicchio blend meant to be drunk the summer following the harvest.

NATALINO CROGNALETTI OF FATTORIA SAN LORENZO

Thirty kilometers from the Adriatic, on the hills of San Marcello, sits the commune of San Lorenzo in Campo. There, on the banks of the Esino River, third-generation winemaker Natalino Crognaletti lives and works on a small farm and vineyard called Fattoria San Lorenzo. An early adopter of organic and biodynamic winemaking, Crognaletti has long relied on simply expressing the natural strengths of this grape on his soil. His commitment to proving Verdicchio's age-worthiness is showcased—and rewarded—by wines like Il San Lorenzo, a 100 percent Verdicchio, rested on lees and aged for nine years.

FABIO MARCHIONNI OF COLLESTEFANO

The Valle Camertina, in Castelli di Matelica, is the only valley completely protected from the sea and it is here, at 450 meters of elevation, that Fabio Marchionni produces what has to be one of the best bottles of Verdicchio in the world. Marchionni, whose grandparents purchased the land, applies the knowledge gleaned during his studies of oenology—he wrote his thesis on Verdicchio—to traffic the crisp minerality from the soil through organic farming, hand-harvesting, and gentle pressing into his outstanding single varietals. From his fifteen-hectare vineyards, Marchionni produces both that standout Verdicchio di Matelica and an extra-brut spumante with 100 percent Verdicchio. From his neighbor's land, he produces an 80 percent Sangiovese/20 percent Cabernet rosé.

More Exceptional Producers
in *Le Marche*

Alberto Quacquarini
Aldo di Giacomi
Andrea Felici
Belisario
Bisci
Bucci
Ca' Liptra
Colli di Serrapetrona
Conti di Buscareto
Garofoli
La Marca di San Michele
Le Terrazze
Lucchetti
Pantaleone
Paris Rocchi
Pievalta
Sartarelli
Stefano Mancinelli
Tenuta Cocci Grifoni
Villa Malacari

Tuscany

Toscana

I landed in Florence at the age of nineteen, having grown up among blocks built in the last seventy-five years in a borough that was farmland a century ago in a country not even 250 years old. The depth of Florence's history—captured in marble and arch; etched onto the banks of the Arno and on bridges across it; captured in tombs, cathedrals, ancient alleys, bronze doors, and the frescoes behind them—resonated in me, with a profound, unforgettable vibration that has kept me bound to Italy ever since. In this, I am far from alone. (In fact, the trope of the young American whose heartstrings are plucked by the beauty of Florence is nearly as old as America itself.)

Florence in particular and Tuscany in general are for many their gateway to Italy and Italian culture. It is the region most often visited by foreign tourists, and it looms so large in the global imagination one might be forgiven for thinking all of Italy is contained in the rolling Tuscan hills and the Duomo-dominated skyline of its capital city.

Tuscany is at the center of Italy not only culturally, but also geographically. The teardrop-shaped region—the fifth largest—spans 8,875 square miles, bordering on Emilia-Romagna, Umbria, and Lazio on its northeastern and southeastern flanks and the Tyrrhenian Sea to the west. Separated in the north by the Apennines and from the south by the Metalliferous Hills and jagged massif of Mount Amiata, its famous gentle hills are fertile, protected, and vast.

The geography of the region has proved auspicious for wine, from the coastal plains of Maremma to the marbled hills of Carrara to the walled cities of Siena, San Gimignano, and Montalcino. Whereas the flat plains of places like Soave in the Veneto have generally proven disadvantageous and the mountains of, say, Trentino–Alto Adige forbiddingly difficult, Tuscany's accommodating morphology seems built for the vine. The climate is temperate, though it does vary from the warmer Mediterranean-kissed west to the cooler Apennine foothills in central Tuscany. The land is two-thirds hilly and one-quarter mountainous, with the rest fertile valley plain.

Chances are, if you're reading this book, you know that, and you're also probably familiar with at least the names of the major players in the Tuscan wine scene: Chianti Classico, Brunello di Montalcino, and Vino Nobile di Montepulciano. All of these appellations rely to a large extent on Sangiovese, which is not only Tuscany's greatest grape but one of the world's best red wine grapes.

What Sangiovese shares with other great Italian grapes like Nebbiolo, Verdicchio, Aglianico, and Fiano is its willingness and ability to express its terroir. The entire apparatus of winemaking rests on that sort of sensitivity as much as, say, the pleasure of listening to a record relies on the sensitivity of the needle. But nowhere does Sangiovese prove itself so excellent an arbiter of terroir as in Tuscany.

The innate quality of Tuscan wines makes the mishegas, or craziness, surrounding their designation, especially in the case of Chianti, so galling. One could fill a *botte* with the ink spilled attempting to illuminate the intricacies of DOC, DOCG, DOCG Riserva, and Gran Selezione, the percentages of what grape is allowed where and in what proportion and why and, since the rules keep changing, when. Generally speaking, these regulations—though perhaps with benign intent—have had a pernicious effect on Tuscany's best-known wine, Chianti. In the 1960s, when up to 30 percent Trebbiano and Malvasia were allowed in Chianti, but not 100 percent Sangiovese, legislation encouraged the mass production of so much inferior Chianti that the image of a straw-covered *fiasco* (flask) filled with barely potable Chianti sitting alongside fiberglass statues of potbellied chefs mid-"Mamma mia" became one of Italy's most depressing cultural exports.

Thankfully, the days when Chianti was synonymous with cheap red-sauce joints are long gone. Of the much-maligned wines of the 1980s—Lambrusco, Soave, Gavi, and Valpolicella—Chianti suffered perhaps most for the misdeeds of rapacious growers. But more recently, as winemakers have evolved and grown more competitive with both each other and international rivals, bureaucratic changes have resulted in endless skirmishes between winemakers and government regulators about what is really Chianti and what isn't. Yet these distinctions have little to do with the quality of the wine itself. Some of the best wines I've had in my life have been Chianti Classico DOCG—as have some of the worst.

But this isn't to suggest that Tuscany doesn't boast some of Italy's premier wines, many of which *do* fall into one or another of the above-mentioned cavalcade of acronyms. However, some of Tuscany's best wines and most promising winemakers have disregarded or rejected the pyramid, preferring to carve their own paths. For this there is ample precedent, especially in Tuscany, where a marquise in the far west of the region revolutionized Italian wine in the 1960s.

For me, Super Tuscans (which, for the record, weren't originally called that by those who first made them) present a real dilemma. When Marchese Mario Incisa della Rocchetta began to commercially produce the first so-called Super

Tuscan, Sassicaia, in 1968 in the town of Bolgheri on the hills of Maremma, he was engaged in an act of rebellion. A scion of one of Italy's noble families, and married to the daughter of another, della Rocchetta longed to turn his vineyards into an Italian Bordeaux. For him, this meant planting Cabernet Sauvignon and Cabernet Franc, as far back as the 1940s, on his *tenuta* (estate) on the Tyrrhenian coast. Della Rocchetta grew Bordeaux grapes in the vineyard, and in the cantina he made wine in a Bordelais style, fermenting it in small oak barrels. After years of experimentation and private consumption, he released the first bottle of Sassicaia in 1968. It was met with immediate success, immense controversy, and a bevy of imitators. In terms of the VVV, Sassicaia has extraordinary terroir and an artisan winemaker, but it lacks one key component: native grapes. Yet it is nevertheless one of the world's great wines. Perhaps more than any other single bottle, it vaulted Tuscany—and Italy—into the same rarefied precincts as Bordeaux and Burgundy. In fact, the 1972 vintage was named the best Bordeaux blend in the world by *Decanter* magazine. For many years, Super Tuscans were so predominant and Sassicaia's imitators so prevalent that in *Bacchus & Me*, the wine writer Jay McInerney felt comfortable saying, "Any Italian wine that ends in the letters *aia* is very good indeed." (Though, it should be said, he immediately follows this with an equally specious piece of wisdom: "There's no such thing as a bad Champagne.")

Della Rocchetta broke new ground, but the harvest from that freshly tilled soil was mixed. Some producers—especially in the monied stretch of Maremma, long the playground for the nobility—leaned more heavily into international varieties. Merlot, Syrah, and Cabernet Sauvignon grew like kudzu, while Sangiovese remained confined to its eastern strongholds. (These western wines with international varieties are the *aia*'s about which McInerney raved.) Other Tuscan winemakers took advantage of their international legitimacy to champion their own Sangiovese. Even today, although Bolgheri was awarded a gerrymandered DOC in 1983 and an even thirstier DOCG Bolgheri Sassicaia, the tradition of disregard remains stronger in Maremma than nearly anywhere else. Not that a general aloofness isn't felt in all of Tuscany. Winemakers like Montevertine, Isole e Olena, Fontodi, and San Giusto a Rentennano have long flouted the strictures of Chianti Classico, preferring to stay true to a 100 percent Sangiovese rather than a blend. More recent additions to outré Chianti producers include Giovanna Morganti of Podere Le Boncie, who makes soulful, natural wine from Chianti labeled, as Sassicaia was, simply Vino di Tavola.

Tuscany is where I first fell in love with Italian wine. And although over the years I've visited nearly every region in Italy, it is in Tuscany and its vibrant, ever-evolving vineyards where I fall in love again.

Native Grapes

VERNACCIA

Vernaccia is another grape whose fate is intimately tied to yet another breathtaking hilltop town in Tuscany. Its home is San Gimignano, a picturesque commune near Siena. As a grape name, Vernaccia is applied liberally to a slew of unrelated grapes, a fact that makes sense only when one realizes that the word simply means indigenous or local in Latin. Every grape is *vernaccia* somewhere.

In San Gimignano, Vernaccia is a very dry white grape, with tempered acid and capable of a delightful saline minerality that completely shape-shifts when aged into something as smooth and flinty as a Sade song. In a region where white wine is often an afterthought—70 percent of Tuscan wines are red—a small group of organic winemakers in San Gimignano are championing their unique native variety, grown near the town since the thirteenth century. For many years, Vernaccia was thought of as *vini turistici*, but, led by Elisabetta Fagiuoli of Montenidoli, winemakers here are committed to giving Vernaccia the care it deserves, hand-harvesting their grapes, and with lower yields. The best Vernaccia is truly unlike any other white wine not only in Tuscany but in all of Italy.

TREBBIANO TOSCANO

It's always seemed to me that Trebbiano Toscano is like a coolly calculated pop song: kinda catchy, pretty insipid, and, unfortunately, unavoidable. No one is in love with Trebbiano Toscano, and yet it's among the most-planted grapes in Italy. Historically in Tuscany, Trebbiano was the creeper in the Chianti, a sort of anodyne parasite to Sangiovese. The grape is sycophantically productive and fantastically dull. Or, it has been up until recently. Lately, for some mercurial reason—and perhaps it's just her iconoclastic impulse—Trentino superstar Elisabetta Foradori has been administering to the much-maligned grape from her vineyard in Maremma. Her wine—Ampeleia Toscano Bianco—is made with 90 percent Trebbiano, with 5 percent each Malvasia and Ansonica. She allows the grapes to be macerated for seven to ten days. The wine's earthy and herbal flavors and bright acidity are a slight but credible reason to reconsider the grape. Another winemaker, Fonterenza, is similarly exploring Trebbiano Toscano's expression in a skin-macerated wine with great success.

Though Foradori and Fonterenza are fighting an uphill battle, Trebbiano has long found at least some success in the niche realms of a Passito-style sweet wine

Giovanna Tiezzi and her husband, Stefano

called Vin Santo, where it's blended with Malvasia. To make the wine, Malvasia Bianca and Trebbiano grapes have traditionally been harvested, then laid on straw mats or in hot winery attics to dry, concentrating the sugars. What little juice is left in the grapes is pressed off about four months after they are harvested and they are put into small casks, called *caratelli*, where they ferment slowly along with the remnant lees of previous batches, or the "mother." These wines age for a minimum of three years, but good producers often wait much longer. The best examples, like those from Fontodi, Fèlsina, Isole e Olena, Castell'in Villa, and Castello di Cacchiano are excellent sipping wines but extremely hard to come by.

ANSONICA

Ansonica is a rare naturally tannic white grape that loves islands. Originally from Sicily—where it is known as Inzolia and is part of the Marsala crew—in Tuscany, it dwells on the tiny islands of Elba and Giglio in the Tyrrhenian Sea. Highly aromatic, the grape carries with it scents of sea air and—perhaps encoded in its genetic makeup—Sicilian citrus. Among my favorite expressions of this intriguing variety are those from Altura, on Giglio; Cecilia, on Elba; and, strangely, Montalcino's Fonterenza. Francesca and Margherita Padovani of Fonterenza make a Toscano Bianco from Ansonica grapes grown organically on the Tuscan coast.

SANGIOVESE

Few knew the name Giulio Gambelli when he died at the age of eighty-six in 2012, but anyone who has noticed a marked improvement in the quality of their Chianti or Brunello or Super Tuscan over the last, say, thirty-five years, has Gambelli to thank. A wine consultant who was born in the small town of Poggibonsi in the province of Siena, Gambelli was the world's foremost expert on Sangiovese, or, as he was called, *il grande maestro di Sangiovese*. Thanks to his resolute advocacy as early as the 1980s and '90s, many winemakers for whom he consulted, including Montevertine, Soldera, Rodano, and La Torre, stayed their hand from adding international grapes to their Chianti Classico. As Gambelli saw it, Sangiovese, though capable of sublime expression, was often overpowered by the presence of showier grapes like Cabernet, Syrah, and Merlot (though it shines in the company of its Tuscan compatriots, Colorino and Canaiolo). Perhaps no one understood Sangiovese better than Gambelli, and yet for years his natural approach— abstaining from the use of commercial yeast, avoiding the domineering new oak, and allowing a bit of oxidation—was dismissed as avant-garde. Perhaps it was. But now, thankfully, his philosophy has been embraced by many Tuscan winemakers, who listen to their grapes with as much solicitude as the late maestro did.

Wonderfully aromatic, with great acidity, capable of aging, and possessing terrific finesse, Sangiovese has many endearing qualities. But its greatest strength lies in its response to the land. But the grape's hardiness has long been at the heart of Tuscany's triumphs and its challenges. When the soil and exposure are ideal, Sangiovese knows no equal. When overcropped, undersunned, or coddled by overly fertile soil, it falls to mediocrity. So, though tedious, much of the hand-wringing over zones is understandable. Because the variety's expression is so varied, it makes more sense to couple grape and place together to discuss three of Italy's most well-known wines.

Chianti Classico

Without falling into the rabbit hole of zones, subzones, and sub-subzones, perhaps the easiest way to describe the Chianti Classico region is as 17,500 square meters of rolling hills between Florence and Siena that are capable of producing high-quality Sangiovese-based wine. (Chianti Classico is part of a much larger Chianti zone, which is of only limited interest to us because of the predominance of industrial producers there.) Today's formula of Chianti Classico was first set forth in 1872 by Baron Bettino Ricasoli, an agriculturalist and, later, Italian prime minister, who stipulated its composition as Sangiovese with a smaller part of Canaiolo, another local variety, to soften the tannic tendencies of the former, and a small percentage of white grapes such as Malvasia and Trebbiano. Today the white grapes are out and Chianti Classico must be at least 80 percent Sangiovese. The other 20 percent can come from more than forty-nine other varieties of red grape. (In practice, though, only a few—including Colorino, Canaiolo, Cabernet, Syrah, and Merlot—are used.) For years, even in Chianti Classico, high-yield low-quality vineyards once predominated. But the struggle for quality wine first appeared as early as the 1920s, when a "Consortium for the defense of Chianti wine and its symbol of origin" was formed to combat cheap imitations. It didn't work. By the late 1970s and early 1980s, Chianti had become a byword for mass-produced wine and its chief grape had fallen into disrepute. However, since the 1990s—when many of the earlier vineyards finally succumbed to overharvesting and had to be replanted—the quality of Chianti Classico has steadily improved. At first this increased promise, led by a cadre of consultants, took form in the addition of international varieties to constitute the 20 percent of non-Sangiovese varieties stipulated. The wines were better than those bottled in the straw-covered *fiaschi* of the mid-century, but they could have been made anywhere. Still, there were always renegades, winemakers like Montevertine, who refused to adulterate their Sangiovese, and in the last twenty years or so, more and more growers have been awakening to the capacity of a 100 percent single-variety Chianti Classico. Others have returned to the classic combination of Sangiovese for character, Canaiolo for comfort, and Colorino for color to make outstanding Chianti.

Along with the realization that Sangiovese had something special to say about the hills of the Chianti Classico region, winemakers have grown increasingly sensitive in both the field and the cantina. Farmers are leaning toward low-yield, high-density organic viticulture. In Panzano, a *frazione* (hamlet) in the northern part of the zone, a group of farmers banded together to create an organic district in which the usage of herbicides is prohibited, even by maintenance crews alongside the roads.

All of this has been to the benefit of the quality of wine from Chianti Classico, but it has made describing it much more difficult. Now that the terroir is more faithfully expressed, with such a variable grape in such varied terrain, generalizations are nearly impossible. As a rule, however, Sangiovese in Chianti Classico tends to have higher acid and lower alcohol than in its neighbor Brunello di Montalcino.

Historically, Chianti Classico has been the coldest of the three regions, and as climate change forces winemakers to higher elevations and east-facing slopes, the wines have gotten more pronounced in their lightness. Giovanni Manetti of Fontodi, for example, has claimed a vineyard at 2,000 feet above sea level called Filetta di Lamole, from which he makes some of the prettiest, softest wines in Chianti. At Val delle Corti, Roberto Bianchi's vines at 1,800 feet yield surprisingly light and minerally expressions of the grape. But one needn't run to the top of the hills to get a great Chianti Classico. Thanks to the tireless work of mostly small producers, the region has been redeemed.

Brunello di Montalcino

Along with Barolo and Barbaresco, the hilltop town of Montalcino and the hills surrounding it, some twenty miles south of Siena, are widely considered at the pinnacle of Italian winemaking. The Sangiovese that grows here is called either Sangiovese Grosso or Brunello. Its small berries and thicker skin yield a darker, more tannic wine than its cousin in Chianti, yet this musculature is also blessed with crystalline definition.

For Sangiovese, there's no better muse than the changeable mosaic that is Montalcino. The hills in the north are of parsimonious limestone-rich *galestro* (shale) soil, yielding more structured wines. The hills to the south are clay, yielding wines of fuller body, but both north or south, most winemakers situate themselves mid-hillside, where the drainage is ideal and breezes caress the vines. The entire area is shielded from nature's extremes by Mount Amiata in the south. But, as is the case around the world, global warming has forced some winemakers to look beyond their mid-hill southern-facing vineyards for parcels with less exposition and higher altitude.

During the peak of Chianti's populist degradation, Brunello di Montalcino quietly watched from the hillside and kept schtum. Mostly the domain of a single family—the Biondi-Santi—until well after World War II, Brunello still enjoys a rather more prestigious spot in the world's cellars. Another reason for its vaunted status is that Brunello di Montalcino has always been 100 percent Sangiovese (despite a major scandal in 2008, dubbed Brunellogate, involving some perfidious producers who illegally introduced French varieties). Importantly, from its inception, Brunello has been known to be worthy of aging. In fact, according to regulations, Brunello wines must be aged for five years before release, including two years in oak and four months in bottle. Barrique was once common, but these days more and more producers are aging their Brunello in either large neutral *botti* or older oak in order to not get in the way of the Sangiovese.

At its best, Brunello di Montalcino can be an expressive, full-bodied, and complex wine, full of red cherry, leather, and tobacco flavors. Though the wines are rich in color, they should not be too dark. The rule is that you should be able to read the pink-tinged *Gazzetta dello Sport* through a glass of Brunello. The fullness of a Brunello is balanced with a good amount of acidity. A Brunello should be like an NFL tight end: big, brawny, yet agile and athletic. Since they can be rather tannic—thanks to the high ratio of skin to juice— they benefit from even more age than required by law. To my taste, ten years, when the fruits are still extant, is best.

Brunello di Montalcino was born into luxury, and in luxury it has stayed. One doesn't venture into the vineyards of Sangiovese Grosso looking for value. Time and tradition have rightly exalted Brunello di Montalcino as one of Italy's signal wines. Bursting with character, with a Titan's confidence, these are wines that feel their full power. Even the bottles would agree they're full of themselves.

Thankfully, the recent tendency toward lighter reds is felt in Brunello as well, and that is showcased in two ways. Within Brunello di Montalcino itself, many winemakers like Stella di Campalto and Le Ragnaie are producing more approachable wines, lighter in body with less, or even no, time in barrique. These upper registers of Brunello are most welcome. But even they are not immune from the relatively long aging requirement and so, to slake the thirst for Sangiovese Grosso's expression as a young grape, a relatively new style, Rosso di Montalcino, has developed.

Rosso di Montalcino wines are essentially the lambs to Brunello's mutton, the veal to its *manzo*. Younger, sprightly, and aged for just a year (at minimum), they often emerge from the same wineries as Brunello but are released into the world earlier. As I was in my early days in Florence, they're unencumbered and hopeful, and they are easy to drink and not exorbitant in price. Among my favorites are Il Paradiso di Manfredi's, Fonterenza's "Alberello" bottling from bush-trained vines, and those of La Torre and Il Colle and the high-elevation vineyards of Le Ragnaie.

More than any other wine region in Tuscany, Brunello di Montalcino is in flux. En masse, virtually the entire old guard has either stepped back or, sadly, passed away. And during the interregnum, the future of Brunello is up for grabs. Among the myriad examples are the sale of the legendary producer Poggio di Sotto, known for its high-toned delicate Brunello, to a larger producer, ColleMassari, in 2011; the passing of Franco Biondi-Santi in 2013; the purchase of traditionalist Cerbaiona by the American venture capitalist Gary Rieschel in 2015; and the death in 2019 of iconoclastic winemaker Gianfranco Soldera, who had left the Montalcino appellation a few years earlier.

The shake-up at the top has been accompanied by a groundswell of young winemakers, leaning, as the young do, toward naturalism in their wines. The leaders of this movement include producers like Stella di Campalto, Pian dell'Orino, and Fonterenza, all of whom are putting their own stamp on these wines, working organically in their vineyards and showing that the terroir of Brunello di Montalcino is one worth celebrating as unadulterated as possible.

Vino Nobile di Montepulciano

In a province where sometimes it feels as if every other hilltop boasts an impossibly romantic medieval town with narrow cobblestone streets flowing into a large piazza where there is a beautiful church and, inside that, some breathtaking Renaissance art and then, around the whole thing, outside the ancient walls of the town, flows a calm sea of vineyards, Montepulciano still stands out. Sixty kilometers southeast of Siena, though it's long been associated with Florence—under whose rule it flourished most fruitfully during the Renaissance—Montepulciano sits atop a 600-meter-high limestone ridge. The surrounding soil is a mix of clay and sand. The slopes are gentle, and the weather is warm. The Sangiovese here, called Prugnolo Gentile, rests between the angular elegance of Chianti Classico and the muscular presence of Brunello di Montalcino. If Chianti is Raffaello and Brunello is Michelangelo, Vino Nobile is Giotto: soft-featured and delicate, yet with underlying strength. This well-balanced temperament has historically endeared wine from here to the many nobles who once called Montepulciano home, ergo the name *vino nobile*. (Well, actually, before 1930, it was called Vino Rosso Scelto di Montepulciano, until a winemaker named Adamo Fanetti thought Nobile sounded better and rechristened it. So, ergo, *the story* of the name.)

But the wine called "the perfect wine" by Pope Paul III's wine steward probably bore little or no resemblance to the modern version. Up until the 1980s, Montepulciano was made by blending Canaiolo, Mammolo, and Trebbiano, with some Gamay. But since the 1980s, Sangiovese has been predominant.

Now, according to DOCG regulations, up to 30 percent non-Sangiovese grapes can be used. And, unfortunately, many producers here fill that balance with French grapes like Cabernet Sauvignon, Merlot, and Syrah. Many, however, don't, and those who rely on Sangiovese alone or in conjunction with Mammolo do the wine justice.

However, at their best Vino Nobile di Montepulciano wines offer an engaging alternative to Chianti Classico and Brunello, often at an extremely enticing value. That can also be the case with the Rosso di Montepulciano, which, like Rosso di Montalcino, is the same wine but a year younger. And many growers in Montepulciano, like Poderi Sanguineto I e II, Salcheto, Tiberini, and Il Conventino, offer both Rosso and Nobile versions that gracefully—and organically—capture the terroir of this beautiful village.

MORELLINO DI SCANSANO

Ever since the rise of Sassicaia, Maremma, the western strip of Tuscany, has been littered with international varieties planted by Mario Incisa della Rocchetta's imitators and hangers-on, the *"aia"* gang. But Maremma is also home to some delightful expressions of Sangiovese, here called Morellino. This is especially true in and around the hill town of Scansano. There the grape is expressed at its most soft and gentle—thanks to the sparse rainfall in the low-lying plains and hills—and is best drunk young. But Morellino di Scansano is also found outside Scansano proper, in the wines of a bevy of biodynamic producers, attracted to the region's hilly western stretches. These include Ampeleia, a collaboration among Elisabetta Foradori, Giovanni Podini, and Thomas Widmann, and Massa Vecchia, a winery located in upper Maremma and run by Francesca Sfondrini. There Sfondrini makes a characterful macerated white based on Vermentino called simply "Ariento," composed of Vermentino, Trementino, and Malvasia Bianca di Candia, and a rosé named "Massa Vecchia Rosato," of Malvasia Nera and Merlot, which another winemaker once described as "not a wine but an emotion."

Giovanna Morganti

Winemakers to Know

GIOVANNA MORGANTI OF PODERE LE BONCIE

It was a bottle of Giovanna Morganti's Podere Le Boncie "Le Trame" that Il Professore tenderly took from a shelf in his Florentine wineshop a decade and a half ago to show me what vino vero was. At the time, I let the syllables wash over me, nearly unintelligible. But the wine I had that day opened the doors to the world I've since inhabited. It was Chianti Classico, I suppose, but nothing like the oaky industrial wine I knew. Earthy, elegant, and intriguing, it was then and is still now perhaps the best expression of Chianti Classico that has ever passed my lips.

Giovanna Morganti is a pioneering Tuscan winemaker who emphasizes high-density, low-yield, hand-harvested, organically grown Sangiovese. Podere Le Boncie is the name of the farm her father purchased in San Felice, a tiny town near Castelnuovo Berardenga in the southern region of Chianti Classico. Podere Le Boncie is also the name of the vineyard, whose vines resemble twisting trees, corralled into line with few grapes visible beneath the profusion and protection of their leaves. And Le Trame—or "The Intrigues"—is a template for so many of the vini veri in this book. It is an expression of Morganti's land, a manifestation of her terroir that refuses to carry any herald other than Podere Le Boncie. In fact, when I first had that bottle, she was part of the DOCG, but in 2012, she parted ways with it. The *consorzio*, the organization responsible for the denomination, rejected her wine, saying it had too little oak and not enough French grapes in the blend. Morganti shrugged and has since been making her wine exactly the way she wants—and the way the grapes demand.

GIOVANNI MANETTI OF FONTODI

Easygoing, warm, and approachable, Giovanni Manetti is the consummate gentleman farmer. From his legendary vineyard in Panzano in Chianti—actually, from the prized amphitheater-shaped valley called the Conca d'Oro (the Golden Snail)—Manetti produces best-in-class Chianti Classico; he is the head of the Consorzio Chianti Classico. But he's also a forward-thinking winemaker with an independent spirit. His 100 percent Sangiovese Super Tuscan Flaccianello is a portrait of a grape in the full confidence of its power and in elegant command. Whenever I taste Flaccianello, I'm instantly transported to Panzano in Chianti, the nearby medieval town where the legendary butcher Dario Cecchini recites Dante over the butcher block, frequently helped along by a glass or two of

Manetti's wine. His Chianti Classico, also 100 percent Sangiovese, is no less elegant while, perhaps, capturing a bit more of the earthiness of the hillside vineyards. His Filetta di Lamole, a wine made at more than 600 meters above sea level, is both a fantastic wine in its own right and an example of how winemakers are adapting to climate change. Though undoubtedly a member of the old guard Chianti Classico winemakers, Manetti has nevertheless enthusiastically embraced organics and biodynamics and even, in small amounts, aging his wines in amphorae.

MICHAEL SCHMELZER OF MONTE BERNARDI

In 2003, the Schmelzer family of Michigan picked up and moved to Chianti. Mother Schmelzer and sister Schmelzer were looking for a change in lifestyle, but Michael Schmelzer, who had trained as a chef, was looking for a new line of work. He found that in the historic vineyards of Monte Bernardi, 350 meters above sea level in the legendary Conca d'Oro in Panzano. Dating back to the eleventh century, the vineyard, with sun-kissed south-facing exposure and

Giovanna Tiezzi

shale-and-sandstone soil, was ideal for Sangiovese. With the zeal of a convert, Schmelzer was intent on making wine that truly tasted of Chianti Classico. Working biodynamically, hand-harvesting, and using spontaneous fermentation but not filtering in the cellar, he released his first wine, called "Retromarcia," or "The Reverse," in 2006. The well-balanced, elegant bottles, which can be drunk upon release but are even better after a decade, sing of the hills of Tuscany.

GIOVANNA TIEZZI AND STEFANO BORSA OF PĀCINA

The patchwork of crops, forests, and vines that dot the sixty-five hectares belonging to Giovanna Tiezzi and Stefano Borsa just south of Siena look, I imagine, almost unchanged from when, in the 900s, an order of monks built their monastery on this land. The Tiezzi family, who have been making wine here since 1967 on land that belonged to Giovanna's father, have similarly been unswayed by time. They have purposefully kept this land undercultivated, at least from a viticultural standpoint. Only eleven hectares are devoted to their vines (Sangiovese, Canaiolo, Ciliegiolo, Trebbiano Toscano, and Chianti Malvasia), fifteen hectares to crops (chickpea, spelt, lentils), and eight hectares to olive trees; the rest remains wooded. Giovanna and Stefano live in the old monastery, which also houses their cantina. Unlike in the somewhat steeper hills of Chianti Classico, Pācina's gentle topography means the vines receive both sun and wind in equal measure and the soil, *tufo di Siena*, balances classic Sangiovese tannins with body and warmth. The farm has been organic since 1980 and chemicals have never been used. "It would be a crime to use chemical treatments here," Giovanna has said. In the cantina, the grapes are similarly respected. Pācina, their most important wine, is spontaneously fermented and macerated in cement tanks, then aged in old oak botti for eighteen to twenty-four months. Their wines are unfiltered. Though they qualify, Giovanna and Stefano have decided against producing Chianti Classico DOCG, preferring to bottle their Pācina, for example, as the more freewheeling Toscana IGT.

STELLA DI CAMPALTO

In the very southeastern corner of Montalcino, one finds the hillside estate of Stella di Campalto, not only one of the region's top producers but, in my opinion, one of Italy's finest. Di Campalto came into this derelict property in 1992 and, moving from Milan to Montalcino with her children, she arrived with a beginner's mind. She had never made wine before. That—augmented by an insatiable curiosity and an innate sensitivity to the land—led her to embrace biodynamic

Stella di Campalto

and organic practices, inspired not only by the philosopher Rudolf Steiner but also by her neighbor, the legendary winemaker Piero Palmucci, of the famed Poggio di Sotto. Working with Sangiovese vines at between 900 and 1000 meters above sea level, she produces Brunellos that are elegant, lifted, and complex, with deep red fruit and engaging softness. Di Campalto doesn't manipulate them much, either in the field or in the cantina. Much of their richness is achieved by her unusually long aging process. She describes her forty-two-month-aged Brunello di Montalcino Riserva as "an egocentric wine with lots to say." It will cost you—her Brunellos hover at more than $100 a bottle—but her Rosso di Montalcino, one of the best reds in Italy, is also among the best values to be had in Tuscany, even if it does go for $70 a bottle.

MARGHERITA AND FRANCESCA PADOVANI OF FONTERENZA

Two more Milanese transplants, the Padovani sisters arrived in Montalcino a few years after di Campalto. Their estate—purchased by their parents in the 1980s, it's on a forested hillside known for the high quality of its water—consists of four hectares, with the vast majority devoted to Sangiovese. The sisters began farming organically and have stayed organic; they planted their vines in 1999, some of which are grown using the labor-intensive *alberello* method, in which vines are grown like little trees, untrained by wire or stake. The Sangiovese-based wines from Fonterenza are lifted with unusual minerality and a focus less on color than on capturing character. Most of their Sangiovese finds itself in a youthful Rosso di Montalcino, though small amounts of Brunello are produced here too. True explorers of the Tuscan vernacular, the Padovani sisters have also been making skin-macerated Vermentino from a few coastal vineyards, as well as various expressions of the much-maligned Trebbiano Toscano.

DORA FORSONI AND PATRIZIA CASTIGLIONI OF PODERI SANGUINETO I E II

With her short hair tucked under a camo hat and a cigarette hanging from her lips, Dora Forsoni is a force of nature. Forsoni and her partner, Patrizia Castiglioni, live and work at Poderi Sanguineto I e II, a small family farm in Montepulciano, 10 percent of which is planted with Prugnolo Gentile, Mammolo, and Canaiolo. It was Forsoni's father, Federico, who planted these vines in the 1960s and who tended them, with Dora, the youngest of nine children, alongside. For years, the Forsonis sold their grapes to *négotiants* for bulk wine, but in 1996, Dora found an old bottle of Sangiovese her father had made for himself, tasted it, and at once realized that her vines were destined for and capable of their own expression. A raspy-voiced hunter (deer trophies hang about the house), Forsoni, a small woman with immense spirit and joy, is inexorably tied to her land. It follows therefore that she eschews intervention of all kinds. The vines are grown organically and the wines are made with no chemicals, fermented in very old oak or in concrete vats with their native yeasts. Every sip of a Sanguineto is a pleasure, not simply because the wines are gentle and themselves noble, but because they recall the indomitable spirit of their wonderful maker.

ELISABETTA FAGIUOLI OF MONTENIDOLI

From Verona, Elisabetta Fagiuoli grew up making Valpolicella at home, not polished wine, but *contadino* wine, with her family. As a young woman, she studied art history; met her life partner, Sergio; and found a stretch of dense wilderness on a hill near San Gimignano. Inspired, she moved there in 1965, had nine children, and from that uncultivated land, began making a transcendent Vernaccia di San Gimignano. Luigi Veronelli described Fagiuoli as a "living metaphor of her land . . . [someone who] knows how to be firm and tough, going against the stream and standing alone but is never unpleasant, never violent." Her wines are the same. For years, she has produced both Sangiovese from the red soil of her land and, from the marine fossil–rich soil, one of the best examples of Vernaccia there is.

FRANCESCA SFONDRINI OF MASSA VECCHIA

Until the 1950s, Maremma, on Tuscany's Tyrrhenian coast, was a mosquito-infested swamp. Dante threw shade on the region as far back as the 1300s, describing it as where "the brute Harpies make their nest." But since the 1960s, it has become one of the more exciting rule-flouting centers of wine production in Tuscany. Today the coast is dappled with high-end hotels and the hills dotted with luxury villas. But it's all earthiness and grounded at the progressive winery on the Metalliferous Hills run by second-generation winemaker Francesca Sfondrini. Founded in 1985, the small six-hectare plot is divided among vines, olive groves, and wheat fields. Sfondrini has tilted the winery toward a more naturalistic style, forgoing sulfur, harvesting by hand, and aging in large neutral *botti* without filtering. La Querciola, made with 80 percent Sangiovese and 20 percent Alicante, is among the best bottlings of Sangiovese in Maremma.

More Exceptional Producers
in Tuscany

BRUNELLO

Canalicchio di Sopra
Cerbaiona
Fuligni
Il Colle
Il Paradiso di Manfredi
La Torre
Le Ragnaie
Montesecondo
Pian dell'Orino
Poggio di Sotto
Salicutti
Sesti
Soldera

CHIANTI CLASSICO

Castell'in Villa
Castello dei Rampolla
Castello di Cacchiano
Isole e Olena
Istine
Monteraponi
Montesecondo
Montevertine
Rodano
San Giusto a Rentennano
Val delle Corti

ISLANDS

Altura
Cecilia

LUCCA

Fabbrica di San Martino
Tenuta di Valgiano

MAREMMA

Ampeleia

SAN GIMIGNANO

Il Colombaio di Santa Chiara
Podere Canneta

VINO NOBILE DI
MONTEPULCIANO

Il Conventino
Salcheto
Tiberini
Villa Sant'Anna

OTHER

Ficomontanino

Umbria

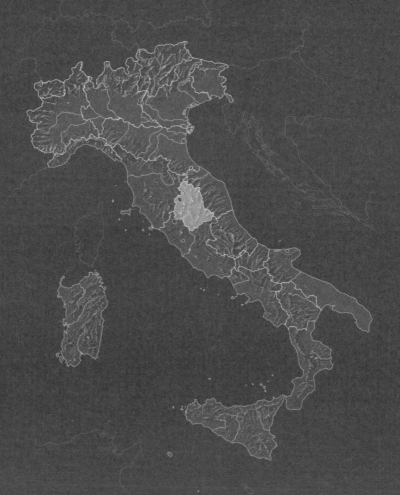

In the fifteenth century, dye makers in Umbria, the landlocked province just south of Tuscany, transformed the manganese-and-iron-oxide-rich soil into the pigment umber. Somewhat somber and foreboding but nevertheless rich and textural, the pigment was used in the dark tones of Caravaggio's masterpieces and the backgrounds of many of Rembrandt's portraits. Some say the name derives from the name of the region; others say it comes from the Latin word *umbra*, for shadow. No matter the origin, in many ways the pigment is a perfect metaphor for a region long overlooked.

For nearly all of its history, Umbria has suffered from comparison to its much larger and more prosperous northern neighbor. (Even dye-wise, sienna is a more popular hue.) Yet Umbria boasts golden hills just as sun-kissed as Tuscany's; towns like Foligno, Spello, Perugia, and Assisi, just as breathtaking as Montepulciano and San Gimignano; and a climate just as mild thanks to the protection of the Apennines, which border the region on three sides. Here it's the Tiber River, not the Po, that runs along the border with Lazio and brings with it warm currents from the Mediterranean.

In terms of wine too, Umbria has fallen under the Tuscan penumbra. Sangiovese, the most widely grown of the native varieties here, doesn't quite reach the heights it does in Tuscany. But Umbrian grapes have their own charms: Among them are the brawny Sagrantino, which grows on the hillsides of Montefalco; Ciliegiolo, the mother grape of Sangiovese; a citrusy white grape from near Spoleto called Trebbiano Spoletino, which is capable of more sublime expression than Trebbiano Toscano; and a real renaissance of Grechetto, a long-derided variety that in Orvieto and its environs yields bright, minerally white wines.

Native Grapes

TREBBIANO SPOLETINO

After centuries spent twenty feet from stardom, Trebbiano Spoletino is ready to step into the spotlight. Long underused for backup harmony in other blends or, even worse, left off the track entirely, this cheerful white grape from outside Perugia is now being vinified as a single variety, emerging with orange-blossom floral notes and a lustrous, almost creamy texture. Its solo debut is a triumph.

As you might surmise from the name, the grape grows in and around the hillside town of Spoleto, as well as in Montefalco. Traditionally farmers grew the grape on the edges of their fields, entwined around the trunks of oak or maple trees, in a rare system called *maritata,* which means married. To give the vines more space, they'd string them from tree to tree like Italian wedding lights. This primitive technique results in ample yields—generally an ill omen for a wine—but Trebbiano Spoletino never lost its acidity or its rich aromatic qualities. Yet it was rarely used for anything but lightening Grechetto.

Among the early champions of Trebbiano Spoletino were the young Stefano Novelli of Cantina Novelli, Giampiero Bea of Paolo Bea, and Giampaolo Tabarrini of Tabarrini, who initiated the grape's ascent in the early 2000s. As testament to the versatility of the grape's talents, each vintner has explored different facets of its character. The most interesting to me are those by winemakers like Bea, Leonardo Bussoletti, and Perticaia, who rely on natural fermentation, because overdetermined vinifying tames the Trebbiano into docility. Today Trebbiano Spoletino is one of Italy's most buzzed-about whites, cresting on a wave of indie cred and creating its own compelling roots music.

GRECHETTO

Pity Grechetto, which for years watched from the sidelines as Procanico dominated the white wines of Orvieto, Umbria's most important white wine region. For decades, Orvieto DOC consisted of vast amounts of Procanico and only scant amounts of Grechetto, Verdello, Malvasia Bianca, and Drupeggio. Sippable but often insipid, it never reached the playoffs. And the whole time, muttering under his breath on the bench, Grechetto fumed, "Hell, Procanico is just another name for Trebbiano Toscano. I can do better than that." The grape was right on both counts. Now, from around this medieval city towering high above the plains atop a volcanic butte, Grechetto is seeing some court time.

Grown both in and outside of Orvieto, the grape has proven itself capable of a range of expressions, from tart and citrusy to full-bodied and appley. In the hands of Francesco Mariani of Raína, it manifests its tannic side in a slightly saline orange wine; in those of Bussoletti, its minerality is at the forefront; and in Tiberi's Il Bianco di Cesare, it emerges as a rustic but excellent orange wine.

SAGRANTINO

Sagrantino is to Umbria what Barolo is to Piedmont or Brunello is to Tuscany: a powerhouse. But what I love about Sagrantino is that this power is expressed not with the brute strength of a bouncer but with the attenuated grace of a dancer. The muscular wine has a tannic, almost feral quality, one reason it was traditionally drunk only as a Passito, a sweet wine. The grape hails from Montefalco, a medieval town with narrow streets that looks out onto quilted valleys. According to Marco Caprai, whose father, Arnaldo Caprai, helped resuscitate the grape in the 1970s, Franciscan monks originally brought Sagrantino to the area in the ninth or tenth century as a sacramental wine. That makes sense, as the word *sagrantino* loosely translates as sacrament. (Later, falconers from Montefalco supposedly salved wounds inflicted by their falcons with the wine.) It wasn't until the 1970s that its dry modalities were explored. It's not an easy wine. However, a few producers, like Paolo Bea, Fongoli, and Milziade Antano, have proven up to the task and produce a Sagrantino that when aged has the burnished strength of an old Hollywood leading man.

More approachable, and immediately drinkable, is the Sangiovese-Sagrantino blend called Montefalco Rosso, which includes up to 70 percent Sangiovese and 15 percent Sagrantino. In this wine, the Sangiovese softens the Sagrantino while the Sagrantino buttresses the Sangiovese. Among the best portrayers of this happy union are Montefalco's Fongoli; Moretti Omero, an organic producer located southwest of Montefalco in Giano dell'Umbria; and, once again, Giampiero Bea at Paolo Bea.

CILIEGIOLO

Long mistaken for Sangiovese, the grape—cherry-like in name and in color—is actually the illustrious child's mother. (Its father, Calabrese di Montenuovo, skedaddled.) The two share both a proclivity for sandy clay soil and the ability to produce cherry-fruited wines. Though Ciliegiolo is grown in other parts of central Italy, Umbrian winemakers have seized upon the grape as their own. At wineries such as that of Leonardo Bussoletti winemakers are creating wonderfully monovarietal wines that burst with succulent fruit, soft texture, and the furious pride of the long overlooked.

Winemakers to Know

GIAMPIERO BEA OF PAOLO BEA

"We don't make wine," says Giampiero Bea, the second-generation winemaker at Paolo Bea. "We generate wine, and I assist in that process." From just five hectares of hillside vineyards just outside Montefalco, Bea coaxes Sagrantino, Sangiovese, and Montepulciano. Then, as delicately as possible, he assists in turning those grapes into wine with such skill that he has become an icon in the natural wine movement and is considered one of Italy's most talented winemakers.

For years Giampiero, an architect by training, worked alongside his father, Paolo, but he took over as winemaker in 1998. That intergenerational transfer twenty years ago was marked by an even deeper commitment to a natural approach to winemaking. Paolo had always abstained from chemicals and additives, but his son went further, working organically and largely biodynamically. Today he's a pioneer in the natural wine movement in Italy. In his cellars, Trebbiano Spoletino rests on its skins, emerging in bottlings like Arboreus, a spiced and baked-apricot-rich wine made from *maritata*-grown grapes, and Lapideus, a zestier version from trellised ones. His individualized natural approach has also resulted in a range of Sagrantino that lacks the strident aggression to which it often falls prey. This he achieves with extended skin maceration, from five to eight weeks, and extended aging, from seven to eight years. In his Pagliaro, named for the hill where the grapes are grown, Sagrantino has a cool, chewy texture, the tannins put into service of the dark plum, cherry fruit, and spice. In his Rosso de Véo (Véo is the name for Bea in the local dialect), which comes from young vines on one of the highest vineyard sites in Montefalco, the grape is bold, elevated, and smooth. And Paolo's Sagrantino di Montefalco Passito, a sweet wine, is aged for ten years and emerges from its oak barrels with notes of coffee, dates, coconut, and salt.

DANILO MARCUCCI OF CONESTABILE DELLA STAFFA

Italy has fostered many big-name wine consultants over the years, and many have advanced an internationalist approach to winemaking that sacrificed individuality for global market share. But Danilo Marcucci, with his white beard, crew cut, and tinted glasses, resembles a hip saint, an apostle for the local, the natural, the VVV. The Umbrian-born Marcucci has worked with many of the region's most interesting wineries, including Collecapretta, Tiberi, and Vini di Giovanni, all of whom he has advised to stay their hands as much as possible. He follows his own advice at Conestabile della Staffa, a three-hundred-year-old estate just outside Perugia. The manor, built after the joining of the Conestabile family of Orvieto and the della Staffas of Perugia in the 1700s (Marcucci's wife is a della Staffa), once included 700 hectares of land, including 100 hectares of vines, and constituted a miniature fiefdom. However, after World War II, much of the land fell into disuse. Today Marcucci tends to just twelve hectares, naturally and organically, from which he produces a fantasia of native grapes—Grechetto, Ciliegiolo, and Sangiovese—in their purest single-variety forms to deliriously vibrant blends.

LEONARDO BUSSOLETTI

Obscure grapes need cheerleaders to pluck them from vineyard obscurity and bottle their magic. Leonardo Bussoletti cottoned on early to Ciliegiolo's potential and, working with researchers at the University of Milan, honed its clones. On his seven-hectare organic vineyard in Narni, a town in southern Umbria whose castle inspired C. S. Lewis's Narnia, Bussoletti and his Ciliegiolo make a happy pair. He produces three different bottlings of the grape, preferring natural fermentation to cultured yeasts. These include my favorite, "Narni Ràmici," and "Narni 05035," which is made without any wood aging and is an agile, sublime expression of the grape.

VITTORIO MATTIOLI OF COLLECAPRETTA

Though the Mattioli family of Colli Martani in southern Umbria has been making wine for more than a thousand years, it is only in the last five or six that these bottles have made it to market. Perhaps this is because the high-elevation, calcium-and-iron-rich soil has produced wine too good to share. Lucky for the Mattiolis; unlucky until now for the rest of us. Thankfully, Vittorio Mattioli has opened up production, making the naturally and biodynamically grown rustic wines available to the rest of the world. (Though, at only 8,000 bottles per year, not the entire world.) Among my favorite expressions from this small family-run vineyard are the amber-colored "Vigna Vecchia," from old Trebbiano Spoletino vines, and the juicy Lautizio, made from Ciliegiolo.

Giampiero and Paolo Bea

More Exceptional Producers in Umbria

Arnaldo Caprai
Cantina Novelli
Fattoria di Milziade Antano
Fattoria Mani di Luna
Fongoli
Montemelino

Moretti Omero
Perticaia
Raína
Tabarrini
Tiberi

Abruzzo

Abruzzi

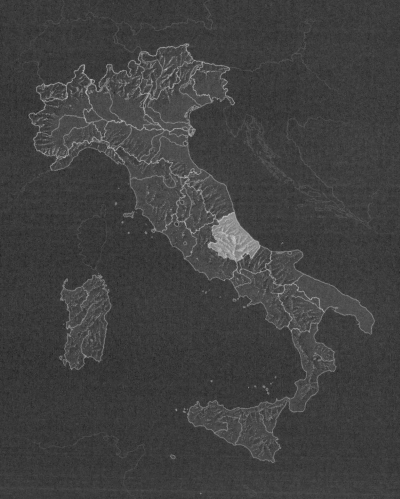

If Tuscany was my introduction to Italy, Abruzzo is my home. I cannot say with certainty why this small region in central Italy has resonated so deeply within me. But love is love, the heart wants what the heart wants, and Abruzzo is my true Italian home.

On my first visit, it was clear to me that Abruzzo, though almost a nonentity in the world of wine, harbored great potential. Its main grape, Montepulciano, has long suffered from an erroneous confusion with Vino Nobile di Montepulciano, to which it is unrelated, and, perhaps, from being a bit too accommodating. Even when prodigiously grown, it retains a soft, fruity joy. (Unlike Sangiovese, for instance, which becomes thin and resentful when overstretched.) It is too nice a grape. I can relate.

In terms of white wine, Abruzzo is home to the outstanding Trebbiano Abruzzese, a sensitive oracle of terroir but too often one that perfidious winemakers have substituted for Trebbiano Toscano, a vapid, if productive, stand-in. In general, Abruzzo has been done poorly by rapacious bulk producers, content to churn out millions of gallons of wines with about as much character as an office building. So although the region is seventh in terms of overall production, it has hardly ever even ranked in terms of quality.

But even a cursory *tour d'horizon* indicates Abruzzo's latent potential. A land of mountains and deep valleys, sparsely populated and largely given over to nature, Abruzzo sits just under the "calf" on the eastern edge of Italy. The Apennine Mountains run north–south through the region, beginning close to the Adriatic Sea and curving gracefully inward as the chain travels south. The best wines in Abruzzo are in the north and the center, in the hurly-burly intermix between the sea and the mountains; on the eastern flanks of the Gran Sasso foothills; and amid the highest peaks of the Apennines. The vines feel the gentle warmth of the sun, there is a diversity of hill soil at their roots, and the sea stretches before them.

Though just east of Rome, and not far below Tuscany, Abruzzo belongs firmly in the cultural ferment of the Italian south. Industry has not the hold here as it does in the northern provinces, and that has had a double-edged effect. On the one hand, much of the natural beauty of Abruzzo remains unspoiled. On the other, many Abruzzians have fled north or abroad seeking economic opportunity since the very first days of the Risorgimento during the nineteenth century, in a flow unstinted by time. Today many mountain towns are languishing, their medieval hearts all but stilled.

Among those who are left, however, is a cadre of generously spirited winemakers with roots so deep in the Abruzzian soil that emigration was never an option. Among these are Emidio Pepe and his family. Perhaps, apart from the landscape and its grapes, what made me fall in love with Abruzzo was this family. When I first visited Pepe and his clan in 2012, having grown up without the pleasing rhythms of family dinners, I was struck by the tight-knit tableau they cut in their villa, Casa Pepe. The undisputed patriarch, eighty-five-year-old Emidio, sat quietly but for a few hushed words. His daughters, Daniela and Sofia, were there too, and Daniela's husband, Giuseppe, an accountant turned chef, was in the kitchen. Daniela and Giuseppe's daughter Chiara, today the global ambassador of the winery, was as bubbly as a spumante but with the depth of a Barolo. Over a salad of tomatoes grown by Emidio himself and pasta made from grain grown on the family's estate, the group argued, laughed, talked, gesticulated wildly, and cracked open old bottles from the estate with so much comfort and ease that I marveled at what unfolded before me, like a kid with his nose pressed up against the glass of a toy store.

Chiara and I are about the same age, and we quickly became friends. One day I felt comfortable enough to ask her why there were so few winemakers in Abruzzo that have reached the quality she and her family have. Was it something about the land itself? "Oh," she said, "well, let me introduce you to another one." That day we drove an hour away to Loreto Aprutino to meet Stefano Papetti Ceroni, an attorney from Bologna turned Abruzzian winemaker who had recently begun making wines under the name De Fermo on his wife's family's property.

Perhaps it was the proximity to the Pepe family and their radiant warmth, perhaps it was because it soon became clear that Stefano and I were fellow travelers in the ways of wine, after one afternoon among the vines, he and I decided to start our own wine label, Annona, specializing in Montepulciano. And through that process, we have become close friends.

Like many Abruzzian growers, Stefano's in-laws had always sold their grapes to one of the forty co-ops in the region, where they were mixed indiscriminately and made into an indistinct wine. With Loreto being situated in an ideal zone for Montepulciano, their grapes, thus used, were like pearls cast before swine. Stefano had befriended winemakers like Elisabetta Foradori and Giovanna Morganti of Podere Le Boncie, and he saw that, if given space and care, his Montepulciano could be a grape of immense charm. When I met him, he had just started his De Fermo label and I was—and am—happy that our wines, grown naturally, have found purchase on some of the best wine lists in America, like those of The NoMad and Charlie Bird in New York and Osteria Mozza in Los Angeles.

Native Grapes

TREBBIANO ABRUZZESE

The relationship between a grape and a grower can be that of an artist and her gallerist. If pushed to overproduce, the market—and often the work—suffers. If the buffeting winds of influence blow too hot, her vision grows blurred and individuality is lost. Trebbiano Abruzzese is Abruzzo's greatest, and most widely planted, white grape. It can be immensely expressive, creating age-worthy white wines full of minerality. (The wines, confusingly, are called Trebbiano d'Abruzzo.) But when yields grow high and alcohol levels hot, its distinctiveness is lost and, sadly, this has often been the case.

Today, however, a new crop of winemakers have dedicated themselves to temperate expressions of Trebbiano Abruzzese. They are led in this endeavor by the mercurial Francesco Valentini, who has been ministering to the sensitive grape since he took over from his father in 2006. In my opinion, Valentini's Trebbiano d'Abruzzo is one of Italy's great white wines, a white capable of aging thirty years or more, with other stellar examples coming from Tiberio's Fonte Canale and, of course, from Emidio Pepe.

PECORINO

As much as a grape that has been grown among the hills of Abruzzo and Le Marche for a century or more can be an up-and-coming grape, Pecorino is one. Long sustenance for sheepherders—hence the name Pecorino, from the Italian *pecora* (sheep)—the grape wasn't expressed as a single variety until the 1980s, when Guido Cocci Grifoni, in Le Marche, and Luigi Cataldi Madonna, in Abruzzo, began experimenting with vinifying it. Pecorino comes from high in the Apennines, and it does well in the cooler vineyards of the mountains of Abruzzo. Pecorino is creamy and herbal in the wines of Emidio Pepe, bright and mouthwatering in those from Torre dei Beati, and saline with stone fruit in Stefano Ceroni's "Don Carlino" Pecorino.

MONTEPULCIANO

No Italian grape has traveled as far in my esteem as Abruzzo's Montepulciano. For years, I had dismissed this red grape as capable only of facile, fruity mass-produced wines that relied on oak in the cantina like a crutch to mask its inherent simplicity. There were, of course, exceptions. Emidio Pepe made a transcendent Montepulciano and Francesco Valentini crafted a rosé called Cerasuolo from the grape, but these pricey bottles were so out of reach financially (for me, at least) they only deepened my conviction that the common swill was more representative of the grape's true nature.

Today, though, it's one of my favorite grapes. As I've discovered, Montepulciano—not to be confused with Vino Nobile di Montepulciano, which is actually Sangiovese—is a grape that rises to meet the care given it and the land on which it is grown. What many had mistaken as Montepulciano's dissolute nature was in fact the product of the large co-ops and industrial producers that occupy southern Abruzzo's flat plains and plant prodigious amounts of the grape. The Montepulciano from the Apennine foothills in the north bears all the characteristics I admire most in a wine: acidity, elegance, age-worthiness, and refined texture, with dark tones. The predominance of middling Montepulciano has to do with the generosity of the grape to give bright fruit flavors even when mistreated. At the higher end of its potential—as illustrated first by Pepe and later by Stefano Ceroni and a host of other producers, including Praesidium and the newcomer Amorotti—Montepulciano can hold serious discourses, rustic earthy backroom talk, and refined exchanges. It's a fluent wine if given the subject matter.

Even more endearing than the grape's dark reds is Cerasuolo d'Abruzzo, a cherry-colored rosato with Montepulciano at its base. For me, these wines—which exist as their own DOC—are among the most exciting rosati in Italy. Though, of course, brighter, friendlier, and more freewheeling than their full-red brethren, Cerasuolo d'Abruzzo rosés retain their structure, at once full-bodied and brightly acidic.

Winemakers to Know

EMIDIO PEPE

Every region, or nearly every region, of Italy has an Emidio Pepe–like character: usually quite old, often elegant, frequently quiet, deeply principled, and devoted to his or her grapes. But only Abruzzo has the actual Emidio Pepe. No one who has met the man forgets him and no one who visits his vineyards in the town of Teramo forgets their visit. As I said earlier, his Trebbiano d'Abruzzo opened my eyes to the ability of Italian white wines to age, and the warmth of the Pepe clan made me fall in love with Abruzzo.

Pepe has been making wine since 1964, but the lineage of knowledge at Casa Pepe stretches back through his father and grandfather to the nineteenth century. Even as a young man, Pepe was laser-focused on returning Montepulciano to glory, or rather, achieving that glory in the first place. Along with Montepulciano, he set his eyes on Trebbiano Abruzzese, which thrived on the clay-and-limestone soil of his fifteen-hectare estate. He never experimented with chemicals or herbicides, not because he wanted to be labeled organic, but simply because of an abundance of respect for the past.

Today that respect continues in the hands of his daughters Daniela and the winemaker Sofia, and his granddaughter, Chiara. The Trebbiano is still pressed by foot. The Montepulciano is stemmed by hand, then fermented and aged in concrete tanks.

Just as he's stayed rooted in the virtuous methods of the past, so has Pepe remained unmoved by market concerns. He follows his gut as to what vintages to release when. Thus, underneath Casa Pepe is a cellar with more than 350,000 bottles, each waiting to pass under the master's eye. Once the appropriate color is reached and the wine deemed ready, the wines are released, resulting in a chronological scramble that doesn't seem to bother the winemakers at all. (Pepe always holds some back.) "The aging cellar is where we have our treasure," confides Chiara. Like their maker, these wines are reserved, complex, principled, and, kept as they are in their subterranean vault, profound.

FRANCESCO VALENTINI OF AZIENDA AGRICOLA VALENTINI

The name Valentini inspires awe among wine lovers. Francesco Valentini, the son of the founder, Edoardo Valentini, who passed away in 2006, grows vines on 80 hectares of the family's 300 hectares of land in Loreto Aprutino. The rest of

Emidio and Chiara Pepe

the three-hundred-year-old estate is devoted to olive oil and cereal grains. Like his father, Francesco abhors publicity: He won't discuss his wines or his methods publicly, won't communicate by email, and doesn't have a website. Nor does he attend fairs, give tastings, or host visitors—in short, the man lives for his grapes. He is so devoted to his wines—of which there are but three: Trebbiano d'Abruzzo, Cerasuolo di Montepulciano, and Montepulciano d'Abruzzo—that when the vintages don't pass his stratospheric standards, he refuses to release them. Even in the good years, the Valentinis bottle only 5 percent of their wine; the rest is sold to locals *sfuso*, or loose, as bulk wine. In his silence, rumors swirl that perhaps his famed Trebbiano d'Abruzzo is in fact made with Bombino Bianco, an obscure grape. But regardless, the wines are aged for years. And Valentini's Cerasuolo, capable of aging twenty years while losing none of its vivacity or complexity, is to my mind the greatest rosé on the planet.

ENZO PASQUALE OF PRAESIDIUM

Prezza, the hilltop village where Enzo Pasquale makes two of the territory's noblest iterations of Montepulciano, was known as Praesidium under the Romans, when it served as a watchtower for the surrounding valley. That strategic importance is echoed in Pasquale's Montepulciano, whose vines grow on five east-facing hectares at 400 meters above sea level. Pasquale, who founded the winery in 1988, has been joined by his son, and they continue to work organically, using fava beans as fertilizer (a technique called *sovescio*) and relying on wild yeasts in the cellar. Apart from their Montepulciano, aged for two years in large barrels but priced as a much younger wine, Praesidium also produces an intriguing dessert wine, a 26.5 percent alcohol-by-volume liqueur made with Montepulciano and Amarena cherries. Its name, "Ratafià," comes from the Latin *ut rata fiat*, which means "Let's sign the thing," and was traditionally used to toast the making of deals.

STEFANO PAPETTI CERONI OF DE FERMO

Of course I'm biased: Stefano Papetti Ceroni is my friend and business partner. But he's also one of the most inspiring winemakers in Abruzzo. In 2007, Stefano, originally from Bologna, and his wife, Eloisa de Fermo, moved to Loreto Aprutino, not far from Valentini's farm, to the country house belonging to Eloisa's family, an estate that included thirty-five hectares divided among vines, olive trees, and grain. As a hobby grower, Stefano stopped using fertilizer and chemicals in 2008. In 2009,

Stefano Papetti Ceroni

he found an abandoned winery buried under the family house dating back to the nineteenth century (it had been active until the mid-1900s) and restored it, and in 2010, he launched De Fermo.

Among the other revelations from De Fermo is that Valentini isn't always completely unreachable. Stefano's stellar Cerasuolo was, in fact, made on the suggestion of the famed recluse. "He told me, 'You are in Loreto Aprutino. You have to do a serious Cerasuolo,'" Stefano said. Fermented in *botti*, his Cerasuolos are among my favorites, but he also produces a concrete-aged conversational Montepulciano and a denser age-worthy expression called Prologo.

More Exceptional Producers in Abruzzo

Amorotti
Cirelli
Iole Rabasco
Tiberio
Torre dei Beati

Molise

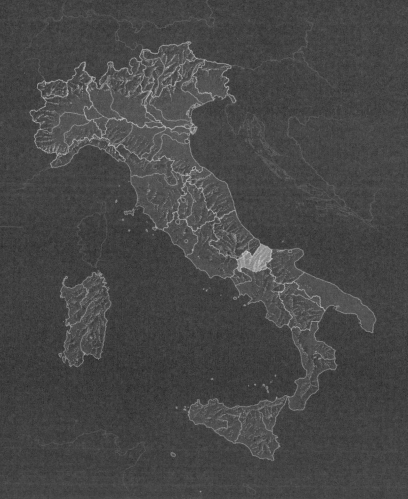

When we opened L'Artusi, the question of what to do with Molise was a doozy. I was committed to representing all of Italy's twenty regions on the list, but I struggled to find any wines being imported from this tiny speck of a region south of Abruzzo, on the country's eastern coast. I wasn't alone in my struggle. When I'd ask Italian friends about the region, they'd often respond, *"Molise non esiste"* (Molise doesn't exist): This was the bemused response with which I was greeted every time I asked after Molise. Little did I know that Molise—and that tongue-in-cheek ontological rebuttal—was something of a national joke in Italy. What started as a hashtag has spun off into groups with names like Molisn't (those who refuse to believe in the region's existence) and Moleasy (those who do believe). Molise does exist, of course. But barely.

For years, the tiny region was aligned with Abruzzo, but in 1963, under still-mysterious circumstances (at least to me), it broke off on its own, becoming a coequal to the mighty Tuscany and the Veneto statutorily, but only on paper. The reality is that this stretch of land, with twenty-two miles of coastline, much of it dotted by fishermen's huts, with the Matese and Sannio Mountains to the west, is sparsely populated and underdeveloped. (In fact, so empty are the villages that in 2019, the regional president offered new residents 25,000 euros to move there.)

What economy there is is largely based on farming, particularly of farro and cicerchia, a slightly toxic but edible green pea. In Roman times and straight through the Renaissance, vines were grown on the well-ventilated hills. After World War II, the ancient vineyards—along with everything else—fell off, only to be replaced by bulk wine producers churning out Montepulciano and Sangiovese, better versions of which are found in both Tuscany and Abruzzo. And yet, as a testament to the democratic and enduring nature of vines, even Molise has a native grape, the Tintilia, and a minyan of winemakers, many of whom are along the Biferno River, are now dedicated to expressing this very existent terroir.

Native Grapes

TINTILIA

Tintilia is a light, aromatic red grape that does well above 400 meters of sea level, but even in ideal conditions, it can be miserly in juice. For those devotees who, through masochism or simply passion, have committed themselves to its cultivation, though, the rewards are a delightfully fruity red wine with an underlying acidity. Drinking a Tintilia is like reclining on a red velvet chaise longue. You are relaxed; you are supported.

As an indicator of just how fragile the existence of lesser varieties can be—and to what extent grapes rely on us to cultivate them—Tintilia wines were well known as recently as 1900, when a Tintilia won the Paris Wine Exhibition. However, the variety had been almost completely forgotten by the mid-1990s, when the producer Claudio Cipressi from Cipressi set about collecting old and forgotten vines from local farmers. Today the parameters of Tintilia are still very much terra incognita, which makes it an exciting grape to follow.

Winemakers to Know

RODOLFO GIANSERRA OF AGRICOLAVINICA

In 2007, Rodolfo Gianserra founded Agricolavinica, 200 hectares of rolling hills in Ripalimosani, of which 30 are planted with vineyards. The rest are forest or devoted to farro, spelt, wheat, orchards, olive groves, vegetable gardens, and pasture. But from his scant 30 hectares, Gianserra grows—organically, sustainably, and with irrigation—an astonishing array of grapes. Of most interest, to me at least, is his Tintilia, which he vinifies in stainless steel into four different iterations: rosata, red, riserva, and Beat, a wine with only 12.5 percent alcohol that is made with carbonic maceration, a technique more common in Beaujolais, which yields a lively and light bottle.

More Exceptional Producers
in Molise

Borgo di Colloredo
Campi Valerio
Claudio Cipressi

Lazio

Latium

Cities breathe life into the countryside surrounding them and the countryside, in turn, provides the cities and their inhabitants with the raw materials for cosmopolitan living. That's true of New York City and the Hudson Valley, and of Los Angeles and the Inland Empire, and it's certainly the case with Lazio and its most famous city, Rome.

After a promising prehistory of winegrowing under the Etruscans, this central western region in Italy fell into millennia of middling production. Romans made up the major market and, as they were seemingly indiscriminate about the wine that appeared in their *osterie* and *trattorie,* wine producers responded by turning hard toward bulk. International varieties like Merlot, Syrah, and Cabernet supplanted native grapes. Co-ops flourished, with each farmer incentivized to produce more. What wines were known beyond Lazio—such as the light white wine Frascati—were anodyne at best.

This has changed in the last twenty years, with vini veri catching fire only recently. The reason, I think, illustrates what both draws us to cities and drives us away from them. On the one hand, the last decade has seen a sea change in terms of wine production. As the ravages of late-market capitalism become ever more apparent, the industrial scale of yore seems increasingly repellent. In its place, the human scale returns. When it comes to wine, that is immensely providential. Though the return to vino vero takes place in the fields, the demand actually starts in urban centers, which have both the market and the mania for great wine. Certainly, that is what I've found over the last twenty years in New York City. Rome is not immune, and as global tastes grew more aligned with the principles of vini veri, the horsepower of Romans' consumer engine began to influence Lazian winemakers.

On the other hand, the same animating principles that induce consumers to seek out vini veri can't help but spur in the heart of many urbanites a desire to escape to the country. So much of the Lazian wine renaissance has to do with the desire of many of us to free ourselves from the frenzied pace of an urban center. Most people accomplish this by drinking vini veri; some accomplish this by *making* it. The last decade has seen scores of young Romans flee to the hills, where, much to their delight, they have discovered that Lazio is prime land for making wine. And their production in turn has spurred Roman restaurants and wine bars like Litro, La Mescita, Rimessa Roscioli, and Bulzoni, to name a few, to focus more on Lazian wines. It's a virtuous vino vero cycle.

These bottles come from a landscape both heir to ancient traditions and young enough to brook rampant experimentation. The raw materials are particularly rich, from the long Tyrrhenian coastline that runs from Tuscany to Campania and provides the region with the sea's warm moderating air to the volcanic soil in the hills on the province's western edge, where the Apennines fade into a whisper.

East of Rome are the Alban Hills, where thirteen picturesque villages are known collectively as the Castelli Romani. From one of them comes Lazio's most well-known wine, Frascati, a white traditionally made with Malvasia di Candia or Malvasia del Lazio (aka Bombino). At its best, it can be bright and crisp, with a treble note of acidity—and no doubt, in the seventeenth century, when it flowed through Roman fountains to celebrate the installation of the new pope, Romans rejoiced. But since the years after World War II, Frascati has tumbled in quality as yields have ballooned. Efforts, often futile, have been made to leverage DOC and DOCG denominations to encourage higher quality, but time is not on Frascati's side. As the Roman suburbs approach and the vineyards slowly retreat, the survivors have increased their yields, thus diluting what had already become a dissolute wine. However, there may be hope yet. Producers like Chiara Bianchi and Daniele Presutti of Cantina Ribelà (the name is Roman dialect for rebirth) have been making Frascati organically from old vines since 2014 (though the wine is labeled Lazio Bianco IGP), and the Carletti family, at Casale Marchese, just thirty kilometers from Rome, in the heart of Castelli Romani, have been making Frascati for the last seven generations.

But those searching the hills for native grapes and high-quality wine need not travel too far. In a small town called Cori, slightly closer to the sea, producers like Marco Carpineti are working with a local white variety called Bellone—his wine is bottled as Capolemole Bianco—and a red grape called Nero Buono, bottled as NZÙ. Farther south and closer to the coast is the town of Priverno, in the valley of the Amaseno River, where two young friends, Arcangelo Galuppi and Emiliano Giorgi, founded Progetto Sete in 2013. Their mission is to rescue forlorn vines from abandoned vineyards. Working naturally with local farmers, they vinify these obscure plantings into bottles of startling originality. Though they do make wines from Bombino Bianco (here called Ottonese), many of their bottles simply go by color. The names of these wines—Freaky, Bomba, Buccia—are of no use at all, but I've yet to be disappointed upon opening one.

As opposed to, say, Piedmont, where land is at such a premium that upstart innovators are more or less confined to the hinterlands (although, as Alto Piemonte shows, *hinter* doesn't mean worse and might mean better), because

Lazio is still undeveloped as wine country, innovative new wineries are larded throughout the region. That said, there are two provinces emerging as particular hotspots: Lago di Bolsena in the north and Frusinate in the west.

In and around Lago di Bolsena, a volcanic crater lake, the predominant grape is an ancient, aromatic red called Aleatico. Traditionally made into a sweet wine, when vinified dry, Aleatico is lightly aromatic, distinctive, and a little bit playful. These hills are filled with career-changers, who far from being dilettantes, arrive in the vineyards with a convert's zeal to express terroir. But it's not all newcomers here either. In Est! Est!! Est!!!—a wine region best known for its emphatic name— Giampiero Bea, of Paolo Bea in Montefalco, has been consulting with an order of Cistercian nuns at Monastero Suore Cistercensi in the town of Vitorchiano. Since the 1960s, the nuns have been farming their small vineyard organically, growing Malvasia, Verdicchio, Grechetto, and Trebbiano. Today Bea makes the wines, including a textured white wine from Trebbiano, Malvasia di Candia, and Verdicchio called Coenobium and an orange wine from the same grapes, called Coenobium Ruscum.

As for Frusinate, though the region was long thought best for white wines, the ascendant grape is now Cesanese, which can make one of Italy's highest-quality red wines. It seems like there's a new producer popping up every month there. No wonder: Cesanese is one of the most exciting grapes in Lazio, red or white.

Of particular note is Marco Marrocco, a former mechanical engineer and elevator company executive turned biodynamic producer and the founder of Palazzo Tronconi. On his eight hectares of vineyard in the town of Arce, Marrocco is dedicated to local grapes so rare they seem like phantoms, with names like Capolongo, Maturano, Lecinaro, Black Ulivello, and Pampanaro.

All in all, Lazio is ripe for development and already on its way there. As the rise of Sicily shows, a region can make great strides in a short period of time. With the engine of Rome to sustain it, I would say Lazio will be truly transformed in ten years' time. For now, there is no coherent narrative beyond better, better yet, and even better.

Native Grapes

CESANESE

Well before Frascati flowed through the fountains of Rome to celebrate the installation of Pope Innocent X in the seventeenth century, the twelfth-century Pope Innocent III was pining for his Cesanese, which he called the "Wine of Kings." The red grape has been grown in Lazio since Roman times but, like so many of the grapes here, suffered in modern ones. The victims of neglect and low-quality co-ops, Cesanese plantings were eclipsed by international varieties. But today, especially in Frusinate, Cesanese is a grape on the rise. Yielding easy-drinking, light-to-medium-bodied, juicy reds with bright fruit and a spicy character, Cesanese is a grape for our times and our tastes.

There are, in fact, three types of Cesanese: Cesanese Comune, Cesanese di Affile, and Cesanese di Castelfranco, though the last exists in such scarce quantities it has taken on the air of legend. Among those three, Cesanese di Affile, with smaller berries and an affinity for higher elevations, is thought to be the best. While in the past these wines were marred by overextraction and over-oaking, today it's the darling of producers wise enough to let the damn grape speak.

This it does from three distinct regions: Cesanese del Piglio, with red iron-rich soils that yield mineral wines; Cesanese di Olevano Romano, with clay soils; and Cesanese di Affile, which also bears *terra rossa*.

Producers like Piero Riccardi and Lorella Reale of Cantine Riccardi Reale make one of my favorite expressions of the grape, called Collepazzo (crazy hills), in which cherry notes and minerality balance as well as a Calder mobile. But Damiano Ciolli, who returned to his family's vineyard in Cesanese di Olevano Romano twenty years ago with the goal of proving that Cesanese could be more than bulk wine, has done just that with two stellar 100 percent Cesanese wines: the delicately perfumed "Silene" and the more complex oak-aged but not oaky "Cirsium." Finally, there's Maria Ernesta Berucci, who has been working to revive her family's sixty-year-old vineyards of Cesanese near Piglio—where the Roman emperor Nerva built a castle just to be closer to his beloved grapes.

ALEATICO

Aleatico, a highly aromatic ancient red grape, grows across central and southern Italy from Tuscany, Apulia, and Le Marche to Sicily, and here in Lazio, where it flourishes on the volcanic shores of Lago di Bolsena in the north. Though it was

traditionally vinified into a sweet wine, a group of producers—led first by Andrea Occhipinti, who set up shop in 2004—have been exploring the grape's dry side, helped by the tempering influence of the lake. Occhipinti, whose Aleatico is soft and aromatic but more savory and mineral than most, has now been joined by Gianmarco Antonuzzi and Clémentine Bouvéron of Le Coste, Joy Kull of La Villana, and Trish Nelson of Vino Gazzetta.

PASSERINA DEL FRUSINATE

Passerina, like Vernaccia, is a family of grapes, some unrelated, that flourishes in Lazio, Le Marche, and Abruzzo. It's generally thought that the Lazian variety—Passerina del Frusinate—has little to do with its Abruzzian and Marchesian cousins. Soft and creamy rather than minerally and acidic (as wines from Abruzzo's and Le Marche's Passerina are), Lazio's wines are generally straw-colored, redolent of stone fruits, and full of flavor and character. Among the winemakers who produce monovarietal bottles of Passerina del Frusinate are Maria Ernesta Berucci and Alberto Giacobbe.

Winemakers to Know

JOY KULL OF LA VILLANA

Nine years ago or so, I sat on the tasting board of the flash-sale site Gilt to help select their wines. Through that process, I got to know Joy Kull, a young woman whose father owned a wineshop in Connecticut. She was, well, a joy, and we became friends. In 2013, Joy, who had been falling deeper and deeper in love with wine, mentioned that she wanted to try her hand at it in Italy. I told her she should check out a winery called Le Coste, just outside Rome. Well, she did. Joy moved to Lazio in 2013, first interning at Le Coste and later finding her own farm in Lago di Bolsena, called La Villana.

Not just because I'm proud to have had some very small and indirect part in their creation, Joy's wines are some of the most exciting and fun in Lazio. Working in a minuscule fifty-square-meter cellar on an old farm, she is making wines from 2.5 hectares scattered around the countryside, bought or rented from retiring *contadini*. Her first vintage, 2016, was made in resin tanks, yielding unfussy, approachable wine. In 2017, she purchased a "cement egg," which allows more transpiration. In 2018, because she loved the effect so much, she bought two more.

At La Villana, Joy explores the range of native varieties and is always trying out new things. She began with a mix of indigenous white grapes including Procanico, Malvasia, Roscetto, and Pettino and a few reds like Aleatico and, of course, Sangiovese, Montepulciano, Ciliegiolo, and Canaiolo. Recently, she planted more than thirty indigenous grapes in her vineyards to see which would flourish in the Lago di Bolsena soil. For now, her wines—called simply La Villana Vino Bianco, Vino Rosato, and Vino Rosso—burst with pleasure and personality. They are structured but not overdetermined, fruity but not florid.

PIERO RICCARDI AND
LORELLA REALE OF CANTINE RICCARDI REALE

Lorella Reale is a Sicilian who studied philosophy and applied ethics in Rome and later became a documentarian, making, among other films, *A History of the Feminist Movement in Italy*. Piero Riccardi was a television director for RAI and author of books such as *The World Upside Down: Two Journalists Traveling Around the Paradoxes of the Globalized World*. The couple were living in Rome when they realized that

Joy Kull

Piero Riccardi

there was more they could do than simply write about the nefarious effects of globalization. They could return to Piero's hometown, Bellegra, forty miles from Rome, to the vineyards of his family, where so many native varieties like Cesanese had been uprooted for French ones, and fight the ravages of modernity by hand. Thus was born Cantine Riccardi Reale.

The couple bought three hectares of Cesanese vines planted around Bellegra and nearby in Olevano Romano. From the beginning, they were committed to biodynamic practices. After three years of experimentation, they released their first vintage in 2013. Both Lorella and Piero are deeply knowledgeable about all aspects of winegrowing and winemaking—she as a sommelier, he as a master pruner—and have proven adept at interpreting their soil. The property consists of two vineyards with varying soils in the area called Collepazzo. Here there's a mix of white sandstone and newer terra-cotta-hued volcanic soil. Both are planted with Cesanese di Affile, but the grapes are bottled in three variations: "Càlitro," from the sandstone; "Neccio," from the volcanic soil; and "Collepazzo," a wonderful balanced example of Cesanese and my favorite of the three.

In addition to these classic dry reds, Cantine Riccardi Reale also produces a lovely rosato called "Tucuca" and an orange wine blend of Malvasia Puntinata and Riesling Renano called "Emotiq."

NICOLA BRENCIAGLIA, DANIELE MANONI, AND MARCO FUCINI OF IL VINCO

Though the grape of the moment is Cesanese, the three friends who founded Il Vinco in 2014 on the southern shores of Lago di Bolsena have devoted themselves to Canaiolo, a red variety grown here for five centuries. Today the grape is mostly known in Tuscany, where it is blended with—and somewhat effaced by—Sangiovese. But on this volcanic soil, its character is revealed as ranging from light, fresh, and juicy to dark, smoky, and brooding, always with alluring spiciness.

The friends—Daniele Manoni, Marco Fucini, and Nicola Brenciaglia—have relied on Andrea Occhipinti, another very exciting producer in Lazio, who lets them make use of his cellar to vinify their five hectares' worth of grapes. Like him, they make their wine organically, with spontaneous fermentation, no additives, and unfiltered and unfined. Though I love their Canaiolo—especially in their Rosso delle Macchie, made from ungrafted vines—I'm most intrigued by the light skin-macerated wine called "Biancoperso," a mix of obscure white varieties including Procanico, Rossetto, and Malvasia Bianca, that is both fresh and savory.

MARIA ERNESTA BERUCCI

When Maria Ernesta Berucci decided to return in 2009 to the small parcels of
Cesanese that remained in her family's possession, she called the wine L'Onda, the
wave. Berucci was, and is, intent on a new wave of Lazian winemaking, and she is
certainly part of it. Located near Piglio, on the slopes of Mount Scalambra called
Colli Santi (meaning holy hills), she and her husband, Geminiano Montecchi, along
with her brother Francesco, have been working together to care for the more than
sixty-year-old vines left behind by her grandfather, Massimi Berucci. (Her father,
Manfredo, once had thirty hectares but transferred all but 1.5 of them to his own
company, Emme Vigneti Massimi Berucci.) In the young Beruccis' hands, these vines
have been trained on pergola, tended to with agro-homeopathic methods (akin to
biodynamics), and turned into stunningly complex expressions of Cesanese di Affile—
rich and brambly, with wild herbs, and spicy, medium in body, and worthy of age.

SERGIO MOTTURA

Not all vineyards in Lazio are near Lago di Bolsena, nor is quality winemaking the
exclusive domain of the young. Sergio Mottura has been working on the Lazian
side of Orvieto, a far northern region that runs into neighboring Umbria, since the
1960s. Though in general, I prefer the Umbrian iterations of Grechetto, Mottura has
long been a champion of the grape. From a long and illustrious line of Piemontesi,
he moved to the area sixty years ago, when land-reform regulations stipulated that
estates must be either occupied by their owner or forfeited. He chose the former
and since then has been working not only with Grechetto, but with Procanico and
Verdello as well. However, it is his Grechetto on which he built his reputation.

GIANMARCO ANTONUZZI AND
CLÉMENTINE BOUVÉRON OF LE COSTE

The eight parcels that make up the fourteen-hectare Le Coste cling to the southern
edge of Lago di Bolsena. Each has its own terroir and its own soil, from basaltic
to volcanic cinder, but they are all administered to by winemakers Gianmarco
Antonuzzi, from Lombardy, and his Lyonnaise wife, Clémentine Bouvéron. The
couple started the winery in 2005 and since the beginning have hewed closely to
natural principles: No chemicals. No herbicides. Their wines are made in a four-
hundred-year-old cellar that belongs to the Antonuzzi family in the heart of
picturesque Gradoli. The wines from Le Coste are a frenetic glorious range from
Bianco R, a skin-macerated Procanico, to the deep-red Le Coste Rosso, made with
Grechetto Rosso.

More Exceptional Producers
in Lazio

Alberto Giacobbe
Andrea Occhipinti
Cantina Ribelà
Casale della Ioria
Casale Marchese
Damiano Ciolli
La Visciola
Marco Carpineti
Monastero Suore Cistercensi
Palazzo Tronconi
Podere Orto
Progetto Sete
Vino Gazzetta

Campania

Fiano. Aglianico. Piedirosso. Greco. Campania, a large region hugging the Tyrrhenian coast, with Lazio above it and Basilicata below, is home to some of Italy's greatest grapes, red or white. Such hyperbole might be expected from a land so dramatic it practically begs for superlatives: The Amalfi Coast, where the Apennines run headlong into the sea, is a scribble of breathtaking beaches indented with grottoes. Naples, the capital, a city of irrepressible energy. Mount Vesuvius, with the trophy of its destruction, Pompeii, frozen before it. The verdant landscape of Cilento in the south, its ancient Roman temples poised between ruin and remain. Campania is as full-throated as an operatic tenor and just as beautiful.

Mountain chains and their foothills course through the region, from the Neapolitan Apennines running along the coast in the west to Cilento and the Lattari Mountains in the south. Pedologically, the region is roughly divided into four sections. Around the Bay of Naples, the volcanic soils of Mount Vesuvius and Campi Flegrei predominate. Between Naples and Benevento, the province to the northeast, lies the Campanian Plain, fertile and under the protection of the surrounding mountains. Due east of Naples is Avellino, with soil that is an auspicious mixture of volcanic rock and porous limestone. In Cilento, in the south, the soil, known as flysch, is a combination of post-volcanic clay and limestone.

A profusion of native grapes is grown in these four regions. Varieties such as Aglianico, Fiano, Greco, Piedirosso, Coda di Volpe, Casavecchia, Pallagrello Nero, Bianco Tintore, and Biancolella mean that winemaking has flourished here for millennia. In fact, Campania is home to the most famous and renowned of all ancient wines, Falernian, from a white grape that grew on the southern slopes of Mount Massico in the north. Today the Falernian tradition is still alive, using Falanghina, another native white.

To some extent, the Vulcans are to thank for Campania's prodigious wine scene. In the caldera of Campi Flegrei, the Phlegraean Fields, thought by the ancients to be the gateway to the underworld, the sandy soil that covers the volcanic rock protected the vines from phylloxera when the disease decimated Italy in the late nineteenth century. Today the soil sustains—though still in its grudging way—rare ungrafted vines of Piedirosso and Falanghina. In Taurasi, the volcanic soil is mixed with calcareous limestone in an area sometimes called the

Barolo of the South. That's nice, but Aglianico deserves to be considered on its own terms. After years spent being made into overly oaky, overpowering, and blustery wines, Aglianico, no longer being pushed to unpleasant extremes, proves itself gracious in its temperate comfort. Luigi Tecce is relying on old vines and a gentle hand in the cantina to produce softer Aglianico. Il Cancelliere, an organic and biodynamic producer in Montemarano, offers up a similar vivid fruitiness. Michele Perillo has climbed to 1,600 feet above sea level to work with old vines that give Aglianico a bracing freshness. Today, Taurasi isn't the Barolo of the South, and Aglianico isn't its Campanian aspirant. Taurasi is Taurasi; Aglianico is Aglianico, and the coming-of-age story of twenty-first-century wine.

Less dramatic but no less striking are the wines of Cilento, a region enveloped by a national park along Campania's southern coast. Though it is not volcanic in origin—it was underwater when the volcanoes erupted—the vines here are exposed nonetheless to sea views and yield the salty yet supple mineral-driven Falanghina and sturdy but silken Aglianico. The park is dotted with independent-minded winemakers, from Casebianche, who makes a small amount of artisan wine, to San Salvatore, running on solar panels, to the godfather of them all, Bruno de Conciliis, a talented and idiosyncratic visionary who has gone so far as to vinify in a cave to protect his precious grapes from electrical currents. He is, without a doubt, one of my favorite wine people.

Overall, compared to that of nearby Basilicata and Calabria, Campania's wine landscape is admirably evolved. That is due to both the benedictions of volcanic activity (no doubt not seen as such at the time of their eruption) and the early interventions of native grape champions. No discussion of Campania would be complete without mention of the Mastroberardino family, who, following the decimating plague of phylloxera in the late nineteenth century chased by the scourge of World War II, tirelessly advocated for the continued cultivation of native grapes like Fiano and Aglianico.

Today the torch of Mastroberardino is carried forward by many Campanian winemakers, including Raffaello Annicchiarico, a Neapolitan microbiologist at Podere Veneri Vecchio near Avellino who makes natural wines in locally produced barrels from native grapes like Greco, Cerreto, Agostinella, and Sciascinoso. Thanks to its natural terrain, Roman tastes, recent advocacy, and contemporary innovation, Campania is near the pinnacle of the vino vero renaissance.

Native Grapes

FIANO

These pages are filled with stories of winemakers who have combed the countryside over the last decade or so, buying old vines and striving to rescue nearly extinct varieties. The story of Fiano is a hopeful model of what those varieties might look like in five or six years' time. An ancient grape, Fiano was in fact nearly extinct when Alberto Mastroberardino traversed the hills of Avellino looking for the last few vines. His first vintage, in 1945, yielded a total of thirty bottles. Slowly—very slowly—the fame of Fiano grew, carried no doubt by the wine's unusual minerality and delicate fruit. The grape, like Assyrtiko in Santorini or Carricante in Sicily, is highly expressive, and its loftiest expression frequently comes from the volcanic and limestone soil surrounding Avellino in Irpinia, in Campania's northeast. For what it's worth, Fiano di Avellino was elevated to a DOCG in 2003. When young, Fiano shows itself capable of being both athletic and lithe; when aged, it is nonetheless fresh but slightly heftier in body, with well-developed almond notes. Among my favorite expressions are those of Ciro Picariello and Cantina del Barone, along with those from Pietracupa, Colli di Lapio, Guido Marsella, and Quintodecimo. It is one of Italy's most exciting white grapes, or, more precisely, the Italian white grape that excites me the most.

GRECO

Greco is a long-praised white grape that grows best, like Aglianico and Fiano, on the volcanic hills inland of Mount Vesuvius. These hills are the home of Greco di Tufo, Campania's most well-known wine. (*Tufo* here refers to the town, not the soil, though the soil is also *tufo*, or tuff.) When planted in high altitudes, the grape finds outstanding minerality and intensity, and even a little smokiness. Among the best versions of Greco di Tufo are those from Pietracupa in Greco di Tufo and from Cantine dell'Angelo, Benito Ferrara, Quintodecimo, and Ciro Picariello.

FALANGHINA

Made from an ancient white grape hailing from the province of Benevento, Falanghina has long enjoyed its status as the breakout white wine from Campania. As a monovarietal, Falanghina can be savory, salty, and a bit smoky, perhaps even bearing sun-kissed tropical flavors (I'm thinking of Agnanum's Campi

Flegrei Falanghina). It also performs admirably as a team player. Among the best expressions of Falanghina ensembles are the wines of Costa d'Amalfi Bianco, in which it is paired with Biancolella; Penisola Sorrentina Bianco, in which it plays opposite Greco; and, from the volcanic cauldron of Campi Flegrei, Cantine Astroni's Lacryma Christi Bianco, where Falanghina harmonizes with Caprettone.

AGLIANICO

Aglianico is the predominant grape in Campania, particularly around Avellino and its twenty-nine communes, though it grows from Falerno del Massico in the north to Cilento in the south. It prospers most notably in the volcanic soils of the Sannio, just west of Benevento and home to the Aglianico del Taburno designation, and on the volcanic-limestone hills of Taurasi, the other of Campania's two red DOCG appellations. Capable of sublime expressions but vulnerable to overweening winemaker intervention—such as was the case with many producers during the 1990s and 2000s—Aglianico can be to southern Italy what Sangiovese is to the central part of the country or Nebbiolo to northern Italy: the clarion grape of its soil.

PIEDIROSSO

Piedirosso (literally, red foot, after the tawny color of the stems) is a red grape that grows nearly exclusively in Campania, on the Vesuvian Hills and the Campi Flegrei. An ancient grape, name-checked by Pliny the Elder, it accounts for just 1,700 acres under vine to Aglianico's 17,000. The grape is often blended with Aglianico to temper the latter's harsher tones, but it sometimes serves as the base grape, relying on Aglianico to bolster its body. As a monovarietal, Piedirosso presents itself as a lively red wine with unusual herbal notes. The best Piedirossi hale from volcanic soil, which allows the grapes to ripen fully without becoming overly alcoholic. Long a supporting player to Aglianico, Piedirosso has learned how to share a stage. In Contrada Salandra's exuberant Piedirosso, the wine contains small percentages of even rarer grapes like Ricciulella, Marsigliese, and Colagiovanna. Monte di Grazia's Piedirosso is blended with Tintore di Tramonti. As far as 100 percent Piedirosso goes, Agnanum's lightly herbal "Sabbia Vulcanica" and the more complex Campi Flegrei "Pér è Palumm" are standouts, along with the easy-drinking, affordable Piedirosso from La Sibilla, a fifth-generation winery run by the Di Meo family, and the excellent Lacryma Christi del Vesuvio from Cantine Matrone.

Raffaele Moccia

Winemakers to Know

RAFFAELE MOCCIA OF AGNANUM

The Moccia family owes their livelihood to the seemingly inhospitable Agnano crater in the Campi Flegrei. The winery, founded by Gennaro Moccia in 1960 and named after the Roman term for Agnano, is full of extremely old and exquisitely gnarled vines, between sixty and two hundred years old. The volcanic soil that makes up these steep hills has protected the Falanghina that grows there from disease. Today the octogenarian Gennaro is joined by his son, Raffaele, to care for these wizened vines tenderly and with the respect elders deserve. The vines enjoy days so hot that cacti pop up from the soil, so sandy ungrafted vines can fully ripen without becoming too alcoholic. As a result of that and the Moccias' devotion to minimal intervention, their Falanghina—with small amounts of Piedirosso—carries crisp character, minerality, and low alcohol levels.

MARCO TINESSA

One of the best wines in Taurasi is made in a garage in Milan. Marco Tinessa, a financial broker born in Sannio but now based in Milan, founded his winery in 2007. The name of his first wine, Ognostro, is Neapolitan dialect for ink, referring to the dark color of this stellar Aglianico, grown on the historic cru Montemarano. Tinessa, who directs the viticulture from the north, trucks the grapes to a garage where he ferments them spontaneously into concentrated but not heavy wines, full of black olive, pepper, and black cherry fruit. His 2018 vintage is the first made in this style, and his wines now include bottles of skin-macerated Fiano also from Montemarano.

CIRO PICARIELLO

Ciro, a man with an endless supply of jokes and anecdotes, is in my opinion the greatest producer of Fiano in Avellino—and, therefore, the greatest producer of Fiano in the world. Working in the town of Summonte, he produces wines free of flaw and full of personality. Though the line includes a *metodo classico* sparkling wine and an entry-level Fiano di Avellino, his 906 bottling is first among equals. So named after the single vineyard from which it is vinified—"#906" on the official Avellino map—the wine is aged a year on lees in steel tanks and then another year and a half in bottles. The high altitude, exquisite care, and inherent nobility of the grape yield a mineral-rich wine that ages and ages and ages.

ALESSANDRO AND ANTONELLA LONARDO OF LONARDO

The Lonardos are a learned family of winemakers based in Taurasi. The winery was founded by Alessandro Lonardo, a retired professor of literature. It is now owned by Enza, his daughter and a biotechnology researcher in Naples, and maintained by Antonella Lonardo, Enza's sister, an archaeologist and literature professor, and her husband, Flavio Castaldo, also an archaeologist and professor of literature.

However, despite the all-star panel of academic consultants, this isn't heady stuff. The Lonardos hand-harvest eighty-year-old Aglianico vines (many ungrafted), farm organically, macerate generously, and produce finely tannic Taurasi wines. Though they are traditionalists to the core, they are not beyond innovation. Among the Lonardos' greatest achievements is the rediscovery of a rare grape, Grecomusc', in 1998. Now renamed Roviello Bianco, the grape is harvested from seventy-year-old vines, aged in steel, and then bottled, filtered but not fined.

Ciro Picariello

LUIGI TECCE

Luigi Tecce was working as a legislative assistant in Rome when his father passed away unexpectedly in 1997. Both grief and a filial sense of duty drew Tecce back to the high southern valley of Taurasi and the Aglianico his father grew there. Unschooled in wine and certain only of the necessity to make it, he began vinifying from his family's eighty-year-old vines high up on the mountainside. Tecce is guided by instinct, but his approach has not failed him. Harvested always by hand, fermented spontaneously, aged both in large chestnut barrels called *tini* and in oak *botti*, his Aglianico has become some of the most sought after from Campania.

GIOVANNI ASCIONE OF NANNI COPÈ

In the eyes of Giovanni Ascione, the nearly extinct Pallagrello Nero could be the next Barolo. Ascione's enthusiasm is infectious, and there can be no quibbling that the twenty-to-thirty-year-old vines he grows on the northwest-facing side of Castel Campagnano, in Caserta, are alarmingly flavorful. Touched with Aglianico and Casavecchia, his Pallagrello Nero displays silky tannins and concentrated flavor, with a pleasingly moderate alcohol level.

PASQUALE CENATIEMPO AND FEDERICA PREDONI OF CENATIEMPO

Ischia, a small volcanic island off the coast of Naples, is frequently bypassed by the crowds that engorge Capri. But Ischia is home to some wonderful wineries like Cenatiempo, run by Pasquale Cenatiempo. Pasquale's father, Francesco, planted the native grapes Biancolella and Forastera in 1945; now Pasquale and his partner, Federica Predoni, organically tend the vineyards 600 meters above sea level.

SABINO LOFFREDO OF PIETRACUPA

Sabino Loffredo, a former gym teacher, lives in Montefredane, outside Avellino. After an injury forced him to abandon his dreams of being a star athlete, he rejoined his father, Peppino, in tending to the family's small vineyard. Soon after taking over in 1999, Sabino decided there was no reason Montefredane couldn't compete with the other wine meccas of Campania. So he began exhorting his Fiano, Greco, and Aglianico to ever-greater heights with immense care. Although such a feat is almost unheard of for an upstart maker, in 2006, his Greco won the prestigious Gambero Rosso Tre Bicchieri award for best white wine in Italy, and I find his Fiano di Avellino just as rewarding a wine.

More Exceptional Producers
in Campania

AMALFI COAST

Marisa Cuomo

AVELLINO

Cantina del Barone
Colli di Lapio
Guido Marsella
Quintodecimo

BENEVENTO

La Rivolta
Podere Veneri Vecchio

CAMPI FLEGREI

Cantine Astroni
Contrada Salandra
La Sibilla

FALERNO DEL
MASSICO

Masseria Felicia

GRECO DI TUFO

Benito Ferrara
Cantine dell'Angelo

ISCHIA

Casa d'Ambra

LACRYMA CHRISTI
DEL VESUVIO

Cantine Olivella

PAESTUM

Bruno de Conciliis
Casebianche
San Salvatore

TAURASI

Guastaferro
Il Cancelliere
Mastroberardino
Perillo

OTHER

Monte di Grazia

Basilicata

Lucania

Basilicata, which sits in the metatarsals of Italy's foot, is rich in natural beauty but poor in resources, infrastructure, and people. The third-least-populated region, it is ahead only of Valle d'Aosta in terms of population density. It is the size of Abruzzo but with only half the people. Cut off not just from Campania in the west, Calabria in the south, and Apulia in the east, but really from the whole of Italy and the rest of the world, Basilicata has the air of the undiscovered about it.

That fact is both celebrated and mourned by Basilicatans. On the one hand, the deliriously mountainous terrain, the dense forests, the spectacular beaches where the region touches both the Tyrrhenian and Ionian Seas, the small villages at great heights, the stone city of Matera, and the pristine lakes are theirs alone. On the other, this poor region is in desperate need of development. The roads are teeth-chatteringly narrow and wind more than a wristwatch. There are no high-speed trains and even few slow ones. No airport either. The novelist Carlo Levi, writing of Basilicata in 1945, stated, "To this shadowy land, that knows neither sin nor redemption from sin, where evil is not moral but is only the pain residing forever in earthly things, Christ did not come. Christ stopped at Eboli." Eboli is in Campania, just short of Basilicata.

When it comes to wine, then, there's plenty of room to develop. For years, it was nearly impossible to find a good wine from Basilicata on a wine list anywhere in America. Now it is just exceedingly difficult. As winemakers throughout Italy fan out across the countryside into regions not their own, a few have found their way to Basilicata. They join a small cadre of Basilicatan winemakers, producing an infinitesimal percentage (0.2 percent) of Italy's wines. Unlike in early every other region, there is no native grape that flourishes *only* in Basilicata. Instead, the most common and important grape is Aglianico, and the best Aglianico grows on the slopes of Mount Vulture, in the north. Without giving too much attention to the wine pyramid, which has not served Basilicata particularly well, these slopes are home to the region's only DOCG wine: Aglianico del Vulture DOCG.

In some ways, Aglianico is the perfect grape for this neglected region and the high volcanic slopes of the mountain its ideal home. The paucity of nutrients in the soil forces the grapes to work hard, to dig deep. Days of brilliant sun caressed by warm sirocco winds give way to astoundingly cold nights. And the result is grapes full of fruit, earth, floral notes, and a volcanic smokiness:

hardship transubstantiated into virtue. In general, I've found the Aglianico from Basilicata to be more rustic than that of its northern neighbor Campania and, unfortunately, at times, still overly oaked. But as new producers align themselves with the principles of vino vero—and old producers like Camerlengo tilt toward the organic—these wines have carried the beauty of Basilicata, like messages in a bottle, across the world.

Winemakers to Know

LORENZO AND ANDREA PICCIN OF GRIFALCO

The logo of Grifalco is a chimerical creature—half griffon, half vulture—and an apt embodiment of this young winery's unique history. After twenty years of making Vino Nobile di Montepulciano in Tuscany, in 2004, Fabrizio and Cecilia Piccin pulled up stakes and headed south to Mount Vulture. The griffon is a nod to the symbol of Montepulciano, the vulture homage to the volcano on whose slopes the Piccins grow twenty-six hectares of Aglianico.

Today the business is run by the couple's sons: Lorenzo, the elder at thirty, in the vineyards; Andrea, twenty-six, in the office. The Aglianico here is grown organically from four separate parcels of land that range from 450 to 580 meters above sea level, with some vines more than eighty years old. The grape appears in four bottlings—an entry-level Gricos, a flagship Grifalco, and two cru bottlings: Damaschito and Daginestra, from the highest vineyards. In all cases, the wines are clean with bright fruit and softer tannins than one might expect, rare for this grape. I find them to be reliably delicious and a great value, walking the fine line between crowd-pleasing and quite distinctive.

JEFF PORTER AND SHELLEY LINDGREN OF VIGNI IGNI

That from all of Italy, Jeff Porter, a sommelier from New York City, and Shelley Lindgren, a restaurateur from San Francisco, chose Basilicata to launch their new wine project, Vigni Igni, bodes well for the future of the region's wine scene. Porter is the former wine director for Joe Bastianich and Mario Batali's restaurants and now host of an online show called *Sip Trip*; he's also involved in many other wine-related

projects, including a wine import company called Volcanic Selections. He and Lindgren, who, as owner of A16 and SPQR in San Francisco, is to Basilicata what the noted restaurateur Bobby Stuckey is to Friuli Venezia Giulia, have joined forces with the Piccin brothers of Grifalco to celebrate volcanic wines (*vigni* means vines, *igni* means igneous). In this project, the pair works with the Piccin family as they vinify from Grifalco's Maschito and Forenza parcels, preferring not Slavonian oak—as the Piccin boys do—but stainless steel. The results are elegant versions of Aglianico with ameliorated tannins but no corresponding loss of distinction.

ELISABETTA MUSTO CARMELITANO OF MUSTO CARMELITANO

Basilicata has long suffered from population decline, but the story of Elisabetta Musto Carmelitano is a hopeful one. This third-generation winemaker was born in 1981 in the United Kingdom, to which her father, Francesco, had emigrated in 1977. But when it came time to build a life, Carmelitano chose to travel to Maschito in Basilicata, on the southern slopes of Mount Vulture, where her father had once cultivated (and then abandoned) sixteen hectares of vineyards and olive orchards. Today, on four of those hectares, Carmelitano—along with her brother, Luigi, and her father, whom she's brought back into the fold—produces three certified organic bottlings of Aglianico from vines ranging in age from thirty (Maschitano) to fifty (Serra del Prete) to ninety years old (Pian del Moro). These wines, naturally fermented, have structure and grace, red fruit, and minerality, like a dusty old-school Sangiovese. Though the total output of the winery is a mere 2,500 cases, Carmelitano also makes a unique Aglianico rosato and even a sparkling white wine in the *metodo ancestrale* from the local Moscato Bianco grapes called "Dhjetë."

ANTONIO CASCARANO OF CAMERLENGO

It takes true devotion and a little madness to focus one's life's work on making wine in a somewhat obscure region with even more obscure grapes. But the architect turned winemaker Antonio Cascarano of Camerlengo hasn't limited his mad devotion to the deserving Aglianico he grows organically in the town of Rapolla. No, Cascarano has expanded his circle of regard to include rare white grapes like Malvasia di Rapolla, Santa Sofia, and Cinguli, the local variant of Trebbiano Toscano. These come alive as vivacious orange wine in a bottling called "Accamilla," which, like all Camerlengo's offerings, is naturally fermented without temperature control, filtering, or fining. "Accamilla" is aged in chestnut and steel; his Aglianico is in chestnut vats, and some finished in oak casks, but in no case does Cascarano intrude upon Aglianico's volcanic smokiness.

More Exceptional Producers
in Basilicata

Azienda Agricola San Martino
Cantine del Notaio
Elena Fucci
Macarico
Paternoster

Antonio Cascarano

Apulia

Puglia

The stiletto heel of Italy, Apulia is an endlessly sunny, extravagantly fertile land framed by the sea, with beaches lining its shores and farmhouses and famous stone *trulli* studding its hills. It's no wonder that it has become the hottest tourist destination in Italy. It was first settled in the eighth century BC by the Messapians, a tribe from what are now called the Balkans, but then, like so much of southern Italy, was traded back and forth (at great cost to its inhabitants) by the Greeks, Romans, Goths, Byzantines, Germans, Saracens, Byzantines (again), and, finally, the Normans, who ruled for seven hundred years, starting in the eleventh century. Then it was back to Apulian ping-pong: the Hohenstaufen, Angevins, Aragonese, Spanish, Austrians, Bourbons, and the French. When Apulia finally became part of Italy in 1860, though, it was far from a liberation. In fact, the Risorgimento opened a long, painful decimation of the land as speculators seized property, condemned Apulians to poverty, and triggered a massive emigration that left the region hollowed out and sucked dry.

That began to change in the mid-twentieth century with the construction of the country's largest aqueduct, which brought irrigation to this dry land— the name Apulia comes from the Latin *a pluvia*, meaning without rain. Today Apulia is overcome by fields, from tobacco in Lecce to tomatoes in Foggia, along with durum wheat, asparagus, turnip greens, olives, and 60 percent of Italy's table grapes.

Perhaps it's surprising then that Apulia has struggled to find its footing in the world of vini veri. In some ways, it suffers from the same problems as central Italy: The vast tracts of relatively flat and fertile land have beckoned industrial winemakers who, with an eye toward profit, plant only the most productive varieties and wring from them millions of gallons of middling wine. This then ends up either as different blends in other parts of Italy or as an indistinguishable and unremarkable table wine. The challenge is compounded by Apulia's position in the very south of Italy, a jut of land all but emptied out by poverty, war, and famine.

It's been a long road back to relevancy, then, one that began in the 1990s with a surge in the popularity of the region's greatest native red grape, Primitivo. These wines—capable of boldness, juicy with ripe fruit, and of good value—swept the international market. But only in the last ten years has the full scope of Apulia's potential started to emerge. Besides Primitivo, there is Negroamaro, which grows on the hot, dry southern tip of the peninsula in the subregion Salento, in an area on a high plateau called Gioia del Colle. From this same limestone soil come some terrific and terrifically undervalued white wines made from little-known varieties like Verdeca, Greco Bianco, and Fiano Minutolo (not to be confused with Campania's Fiano).

More than most, Apulia is known not by its grape varieties but by the color of its wine. As the rest of Italy, save Abruzzo, is just catching on to rosati, Apulia has long been aware of the charms of pink wine. In fact, the region accounts for 40 percent of all rosati made in Italy. Far from the fripperies of the Provençal style, Apulian rosati are of a darker hue and made mostly from Primitivo and Negroamaro. One can't imagine the extremities of Salento giving forth wines with anything but robust character, and the rosati don't disappoint. They're big, generous, and a bit gregarious, with fruit balanced by minerality, intensity tempered with joy. The best, in my opinion, are made with Primitivo and come from the vineyards of Pasquale Petrera of Fatalone, Cantina Pantun in Mottola, and Guttarolo in Gioia del Colle, which are aged in amphorae.

Native Grapes

VERDECA

A white grape long used almost exclusively for the production of vermouth, Verdeca is now being vinified by producers such as Li Veli from the sunny Salento plain, Cristiano Guttarolo at Guttarolo, and Francesco Valentino Dibenedetto at L'Archetipo as a crisp, dry wine. Grown in central Apulia in the provinces of Taranto and Bari, 100 percent Verdeca wines, though rare, can be lean, clean, and saline.

PRIMITIVO

Primitivo is the king of southern Apulia, with its seat in Gioia del Colle. It was to here that a Dalmatian priest brought the grape from neighboring Croatia, naming it Primitivo. Far from indicating any atavistic primitivity, the word comes from the Latin *primativus,* meaning first to ripen. Primitivo is so closely related to Zinfandel that, since 1999, producers have been allowed to label their wine thusly. Like Zinfandel, the eager grape can tend toward very high alcohol and so needs a deft hand in the vineyard and cantina. The reputation of Primitivo, which had long been used as a blending grape, needed rescuing in the 1990s. Now it is generally accepted that the dark, spicy fruit, reined in with acidity, is a worthy grape and capable of great aging. Yet dangers abound. More than most, when pushed by climbing temperatures, Primitivo develops sugars quickly, yielding wines with unbalanced alcohol levels. As climate change brings hotter summers, this will inevitably affect already-scorched Apulia. Yet hopefully, by utilizing organic methods, wise and prudent pruning, and delicate vinification, winemakers will continue to bring the noble grape well-deserved accolades.

NEGROAMARO

Do not let the false cognate deceive you. *Amaro* means bitter in Italian, but the name Negroamaro does not indicate bitterness. It is instead derived from the Greek *mavros*, which, like *niger* in Latin and therefore *negro*, means black. So the name is, at least etymologically speaking, black-black. Not as well known as Primitivo, Negroamaro is more widely grown. One of Italy's most ancient varieties—it was brought by the Greeks in the eighth century BC—the dark-skinned grape offers a similar fruit to that of its brother Primitivo but with softer tannins.

Like Verdicchio in Le Marche, Negroamaro suffered from an ill-conceived EU vine-pulling scheme that resulted in many of the older vines being grubbed up. Once there were 31,000 hectares planted; now there are just under 12,000. The grape flourishes in Salice Salentino, on the eastern edge of the southernmost appellation in Apulia. Historically the grape has been blended with Malvasia Nera, which adds acidity, but of late, winemakers have begun pairing it with Primitivo as well. In the former blend, the wines are spicy and fruity, with notes of cinnamon and clove, cherries and blackberries; in the latter, they tend to show flavors of tobacco, licorice, and leather, along with an earthy minerality.

Negroamaro remains an excellent value for the money, despite the growing interest in this grape. Among my favorites are Guttarolo's Murgia bottling and Natalino Del Prete's chocolaty Rosso Anne. On the juicy, fruity side, Perrini's Negroamaro from Salento is another bottle of great value.

NERO DI TROIA

A century ago, Nero di Troia was widely planted in northern Apulia, in and around the town of Castel del Monte. Today, only 1,800 hectares remain. A powerful grape, Nero di Troia was part of the flow of wine from Apulia into the cellars of French winemakers, who relied on the late-ripening, thick-skinned fruit to add body to their anemic vintages. In Apulia, the grape has largely been blended with Montepulciano, to soften the potentially overpowering tannins resultant from underripe berries. But when fully ripened and allowed a solo turn, Nero di Troia emerges with dark berry notes, gentle tannins, violet aromas, and alluring spiciness.

Winemakers to Know

PASQUALE PETRERA OF FATALONE

Like Negroamaro, Fatalone bears false cognates if incorrectly dissected. Far from being a portmanteau of "fat" and "alone," it means "one who can boast to be an irresistible seducer of women." The name was given to Filippo Petrera, son of the winery's founder, Nicola Petrera, who chose this hill on the eastern side of the Gioia del Colle. Filippo, Il Fatalone, drank a half liter of milk and a half liter of Primitivo every morning and lived to be ninety-eight years old.

Today the current owner, Pasquale Petrera, leads the farm and vineyards (and drinks less). One of the first winemakers to be certified organic, he takes great pains to make his grapes comfortable. He even plays them music—classical with some nature sounds mixed in—in the cantina. But perhaps that is all part of his overall approach, which could be called "Do no harm." The winery is carbon neutral, solar-powered, and 100 percent sustainable. Petrera relies on a long, slow process of spontaneous fermentation.

The first to bottle Gioia del Colle as a single-variety DOC in 1987, Fatalone expanded to include Greco Bianco in 1993. Here the grape is floral and alluring, with both tropical and underripe flavors and a zing of intense minerality. Although Fatalone's 100 percent Primitivo is certainly one of the region's best, my favorite bottling from this estimable cellar is "Teres," a light Primitivo—actually a rosato, though Petrera doesn't call it such—that reminds me of Cerasuolo d'Abruzzo and flies off the list at LaLou, especially during the summer.

CRISTIANO GUTTAROLO OF GUTTAROLO

Working in the stable of an ancient stone farmhouse high on the Murge plateau in Gioia del Colle, Cristiano Guttarolo produces some of the most elegant wines in Apulia. Since starting his winery in 2004, he has been devoted to organic and low-intervention methodology. No tilling. No irrigating. And certainly no chemicals. His vines—Primitivo, primarily, but also some Negroamaro and scant amounts of a rare red called Susumaniello—grow on *carso*, the same limestone soil found in Friuli Venezia Giulia. The relatively high altitude is augmented by the cooling Mediterranean winds sweeping over the vineyard, allowing for late harvests and the development of unusually profound flavors. A relentless pioneer, Gutarrolo was the first to age his Primitivo in amphorae, resulting in 2,000 bottles of highly sought after wine.

NATALINO DEL PRETE

For years, the quiet, gray-haired Del Prete flew under the radar, letting his work on his forty-acre estate in Salice Salentino speak for itself. Here he grows Primitivo, Malvasia, Negroamaro, and Aleatico in grass-filled fields, without fertilizer, plowing, irrigation, or chemicals. He vinifies in a massive cantina that once belonged to a now defunct *négociant*. Del Prete occupies just a fraction of the space, where he produces 2,000 cases of wine every year. His wines course with an earthiness that can only come from *mezzogiorno,* as southern Italy is known, his forty acres, and a life spent tending the vines.

More Exceptional Producers in Apulia

Cantina Pantun
Francesco Marra
L'Archetipo

Li Veli
Perrini
Polvanera

APULIA

Calabria

Calabria is a region whose artisanal winemaking potential has not yet been fully tapped. In fact, sometimes it feels as if its potential has been knocking politely on the door forever, too quietly for anyone to hear. The region—just larger than Campania but with one-fifth the population—makes up the entire ball and toes of Italy's foot, separated from Sicily by the shocking-blue Strait of Messina. Calabria is Italy's end, where the Apennines, dividing the land between the Ionian Sea in the east and the Tyrrhenian in the west, peter out. Calabria is ringed with rugged rocky coastline, with winds that never cease blowing and sun that just won't quit.

Calabria is the poorest region in Italy, one reason why the Calabrian diaspora has been one of the country's largest. Its natural resources are mostly unexploited, except when it comes to the famous citrus—Calabrian oranges, bergamot, and citron are some of the best on the planet, and their groves stretching under the brilliant blue sky form an indelible image—and olive oil, of which the region's production comprises a third of Italy's total, though the vast majority is industrially made and of relatively low quality.

Wine isn't unknown here, of course. When the ancient Greeks colonized the peninsula in the seventh and eighth centuries BC, they referred to the indigenous inhabitants as the Oenotri, meaning people who cultivate grapes. But although the region was home to 180 native varieties, these were decimated by phylloxera in the late nineteenth century, a calamity from which Calabria, at least from a vino vero standpoint, has since struggled to recover. For many years, the only wines that were exported came from Librandi, which produces 2.5 million bottles a year (much of it with the native Calabrian grapes that did survive); the rest was simple table wine meant for local consumption. Many farmers, faced with existential economic hardship, turned to the more profitable olive oil trade. It's only been in the last decade or so that Calabrians have begun to embrace artisanal winemaking, taking up once again the mantle of the Oenotri.

Today there are at least a dozen forward-thinking producers devoted to making wines from native grapes, most notably Gaglioppo and Greco Bianco, and many of them are farming organically. From the raw but expressive wines of Nasciri to the juicy, elegant bottlings from À Vita, the quality is continually improving, and Calabria now offers perhaps the best value of all Italian wines.

A mountainous and hilly region with little pedological consistency, Calabria lacks an overall viticultural narrative. In the eastern corner is Cirò, by far the most well known winemaking area. Here the grape is Gaglioppo, grown on extremely old bush-trained vines, which tumble from the La Sila plateau toward the Ionian Sea. The grape's thick skin, natural protection against the unrelenting sunshine, yields intense rustic tannins, but these are being tamed by À Vita, Du Cropio, and Sergio Arcurì, to produce pale, light-bodied reds with good acidity, a touch of oceanic salinity, and a signature herbal quality. From Cirò's Gaglioppo also come expansive, expressive rosati, full of juice and joy. These are perhaps the most exciting expressions of the grape.

Gaglioppo's white counterpart in Cirò, though a mere cameo in terms of production, is Greco Bianco, which yields a fresh, lively, and acidic wine. Though less mineral than Greco di Tufo, whose upbringing in the volcanic soil of Campania ensures its longevity, Greco Bianco from Cirò is lightly floral, fruity with a glimmer of salinity, and pleasingly mellow. For now it's best drunk young, but, with the fast progression of Calabria's wine world, that may not be the case for too long.

Apart from those in Cirò, Calabria's winemakers are far-flung and disparate. In a wild forested northern sprawl called Pollino, where the Apennines have their last flourish before fading into the sea, is Saraceni, long known for its sweet wines, made from Malvasia Bianco. Today, though, growers like Cantina Maradei and Giuseppe Calabrese have turned to old vines of the local grapes Magliocco Dolce and Guarnaccia Nera to make intriguing dry red, white, and rosati wines.

In the far west, near the town of Pizzo, winemaker Giovanni Benvenuto has devoted himself to a rare local grape called Zibibbo di Pizzo, which he grows from a steep terraced vineyard high above the Tyrrhenian Sea. The grape, nearly extinct, yields floral and aromatic wines, well balanced, shy of cloying, and certainly an argument for its continued propagation.

Also on the western coast, though more southerly, toward the Savuto River, the brothers Battista of Odoardi make a range of wines using local grapes like Gaglioppo, Magliocco Canino, Greco Nero, and Nerello Cappuccio, a grape known better across the Messina Strait in Sicily (its inclusion makes sense when you realize the Aeolian Islands can be seen from their 270-hectare farm). Among my favorites from this winery is Savuto, a suicide soda of Gaglioppo, Magliocco, Aglianico, Greco Nero, and Nerello Cappuccio.

Native Grapes

GAGLIOPPO

Gaglioppo is Calabria's flagship grape. The rich tannic red wines from Cirò are perhaps Calabria's best known, and there's something of a resonance between the thick-skinned grape, a defense against the intensity of the environs, and the thick-skinned, fiery intensity of so many Calabrians. There are certainly stellar examples of these traditional reds—especially by Du Cropio and À Vita—but Gaglioppo expressed as a rosato should not be overlooked.

The short maceration yields lower tannins, yet the resultant rosati are nonetheless deeply textured, juicy, and fun. They are the perfect expression of these hot climes, where life is slow but fermentation is fast. More important, a Gaglioppo rosato is the perfect foil for the furiously spicy Calabrian cuisine, in which the main protagonist is the peperoncino. Among my favorite Gaglioppo rosati are those of Sergio Arcurì and of À Vita, who also make 3,400 quickly moving bottles of an elegant red.

Francesco de Franco

Winemakers to Know

FRANCESCO AND LAURA DE FRANCO OF À VITA

From the hills outside the village of Cirò, the Ionian Sea on one side and the forested Sila Mountains on the other, À Vita's bright and lively wines are made by Francesco de Franco, a former architect from Cirò, and his wife, Laura, a Friulian. The two, who founded the winery in 2003, grow their fifty-year-old *albarello*-trained Gaglioppo, intermixed with fig and olive trees, organically, relying on the calcareous clay and extra time in the bottle to yield intense expression with soft, elegant tannins. The riserva, which is aged for at least three years before release, is dusty and spicy, reminiscent of a cross between Sangiovese, Nerello Mascalese, and an old-school Pinot Noir.

GIUSEPPE IPPOLITO OF DU CROPIO

When I first began my career in Italian wine, I, like many, gave Calabria short shrift. That changed when I met Giuseppe Ippolito. With his snowy-white hair cut short, arched eyebrows, and cherubic features, Ippolito was something like the Hermes of Calabria, or, maybe better, Bacchus. Du Cropio arrived in New York City with not just his wines but also a cornucopia of Calabrese food: spicy, spreadable 'nduja sausage; cheese; chilies; and soppressata di Calabria.

It wasn't just a charm offensive. What Ippolito knew, and I soon came to find out, is that the wines of Calabria must be seen in context and when they are, they're just as flavorful and unique as anything else from Italy.

Ippolito's winery and vineyards are in Cirò Marina, where his family has been a presence for over 160 years. Du Cropio occupies a scant thirty hectares along the steep slopes facing the Ionian Sea. Like the man who makes them, the wines are gregarious, big characters bursting with life. His "Serra Sanguigna," a Gaglioppo-based blend with small amounts of Greco Nero and Malvasia Nera, is immensely drinkable. His Cirò Rosso Riserva "Damis" is extremely concentrated and a great example of an age-worthy 100 percent Gaglioppo, still vibrant despite being aged for seven to eight years before release. My favorite wine from Du Cropio sits between the Serra Sanguigna and Damis and is called "Dom Giuvà." Gaglioppo blended with 5 percent Greco Nero, the wine displays firm tannins, neither as rustic as the Serra Sanguigna nor as impenetrable as the riserva.

DINO BRIGLIO OF L'ACINO

A film director, a historian, and a lawyer walked into a vineyard. . . . Thus begins the story of L'Acino, which means "grape" in the local dialect and is the project started by three friends in 2006. Today it's only the historian, Dino Briglio, who runs the ten hectares of land on the border of Pollino they bought from an old farmer. Briglio grows local varieties, including the red Magliocco Canino and the white Mantonico on a sandy hillside there. The wines, fermented with natural yeasts and aged for the most part in stainless steel, are unfiltered and unfined. I'm particularly partial to the Giramondo, a balanced, clean orange wine.

GIUSEPPE CALABRESE

Another young winemaker near Pollino, Giuseppe Calabrese returned to his grandmother's four hectares of vineyards in 2007 after dropping out of agricultural college. Calabrese is focused on Magliocco, which is, for this northern region, what Gaglioppo is to Cirò, and Guarnaccia, a rare white varietal that yields a savory, well-rounded wine laced with salinity. Since 2013, he's been bottling his own vintages, made without commercial yeast or temperature control and tasting exactly of this relatively unknown pocket of Calabria.

More Exceptional Producers in Calabria

Azienda Agricola Nasciri
Benvenuto
Cantina Maradei
Casa Comerci
Odoardi
Sergio Arcurì

Sardinia

Sardegna

Sardinia is Italy, but only barely. Sardinia is, above all, Sardinia. The second-largest Mediterranean island, behind Sicily, located 190 miles off the coast of Lazio, Sardinia is a region that keeps to itself and keeps itself shrouded in mystery. Apart from the sliver of La Costa Smeralda, The Emerald Coast, the playground for plutocrats that runs along the northeastern edge, Sardinia is largely unknown by non-Sardinians. And that seems to suit it just fine.

Sardinian culture, heavily influenced by four hundred years under the Spanish and by a long procession of ancients before that, is as insular and rich as a geode, as mysterious as the stone *nuraghi*, the granite totems left by Neolithic Sardinians one can still find studded in the hills. Each culture left its own traces, and today Italian is joined by Sassarese and Gallurese, mixes of Corsican and Sardo; Algherese, a form of Catalan found on the west coast; and Tabarchino, a variant of Ligurian, with, of course, Sardo as the official tongue. This semiautonomous region, which is separated by only eleven kilometers from French Corsica, jealously guards its culture. One might think of the colorful headdresses of traditional Sardo costumes, or *cantu a tenore*, one of the oldest examples of polyphony, or the proud tradition of Sardo literature. The island is the birthplace, home, and muse of Grazia Deledda, who in 1926 was the first Italian woman to win the Nobel Prize in Literature, for her stories capturing the folklore-soaked world of Nuoro, a town in central Sardinia. Starting in the 1980s, Sardinian literature has enjoyed a renaissance called the Sardinian Literary Spring. Although many of its authors have not been translated into English, all sing of their country with nuance, skill, and pride. What concerns them, and many Sardinians, is the idea of *sardità*, Sardinianess.

Though Sicily is only slightly larger than Sardinia, it makes six times as much wine. (Sardinia, in fact, exports more wine corks than it does wine.) Instead, these wild landscapes are given over, largely, to pastures for sheep. (As in Wales, the joke is there are more sheep on the island than people, the accuracy of which in no way compromises its truthiness.) What grapes there are in Sardinia tend to grow on small parcels in crags and valleys in the island's western half. The region is full of indigenous varieties that are only beginning to be studied. Research published in 2012 indicated that there are 250 grape varieties grown on the island, of which only 24 have been registered. Among the most well known are Vermentino, Nuragus, Cannonau, Carignano, and Vernaccia di Oristano, a distinct grape from Tuscany's

Vernaccia. (Vernaccia means local.) That Cannonau is Grenache and Carignano is Carignan in France, in my mind, in no way disqualifies them from being considered indigenous, since they've been here for hundreds, if not thousands, of years. Sardinia is also home to varieties just barely on this side of discovered—wine from grapes like Nasco and Semidano, of which only 120 acres are planted, Bovale Sardo, and Pascale rarely make it off the island. These grapes, along with the more properly indigenous Nuragus, have that all-important quality of *sarditá*.

Most vineyards are small, on average only 2.5 acres, which means that Sardinian wine has, for the most part, been made by government-sponsored cooperatives, with the attending erratic quality. Many of these were overly rustic, some too alcoholic, and others too oaky and generally just unbalanced. But over the last twenty years, Sardo winemakers have, like other Italian winemakers, moved to cleaner, more exacting wines. This pertains to both the remaining cooperatives and a cohort of forward-thinking small producers.

Some of the best wines, especially Carignano, come from ungrafted vines. (The island, while not completely spared from phylloxera, was not quite as decimated as most parts of Italy.) While that is not unique to Sardinia, those deep-searching roots are even more important here, where the climate is scorchingly hot and parchingly dry. The vines that do survive are rewarded with the cool mistral winds sweeping down from the Rhône Valley, while those in the north enjoy the moderating effects of the Mediterranean. Regardless of the variety, one can nearly always detect a little of the land's dryness and the wild herbs that grow everywhere on the island. The strength of character of the landscape is distilled into every bottle.

For most regions of Italy, both their resources and potentials are known. In some, the resources are there but their potential has not been fully explored. But Sardinia continues to be terra incognita, an intriguing, alluring wild frontier.

Native Grapes

VERMENTINO

Vermentino grows all over Sardinia, but it is especially suited for Gallura, a region on the northeastern tip of the island, where it enjoys the cleansing breath of the mistral, sweeping south from the Rhône. These intense winds yield aromatic and fruit-forward wines, somewhere between Maremma's ripe fruit and Liguria's more mineral-driven style of Vermentino. And the winds, along with the natural sandy soil, are key to keeping the grapes dry and well ventilated. The results are a uniquely Sardinian profile: a prominent minerality with bitter almond, notes of ripe stone fruit, and a distinct saltiness. Fabrizio Ragnedda's Capichera put Sardinian Vermentino on the map when it released the first monovarietal wine in 1980, but today the grape plays muse to an orchestra of interpreters, from Deperu Holler to Piero Mancini to Vigne Surrau to Alberto Loi.

NURAGUS

Two-thirds of Sardinian wine is white, and the most planted white grape is one of Sardinia's truly indigenous varieties, Nuragus. The grape has been on the island since the Phoenicians arrived in the twelfth century and is known by a variety of names, including Abbondosa, Preni Tineddus, and Axina Scacciadeppidus. All of these—"Abundance," "Grab the buckets," and, my favorite, "The grape that pays off debts"—nod toward the grape's tendency for prodigious growth. And that overabundance has traditionally yielded flabby, unremarkable wines.

However, yields have been dropping as winemakers awaken to the fact that the grape can produce compellingly light, delicate wines with low alcohol but high flavor, a tonic of green apple notes, and pleasant, persevering bitterness. Among the wines that best capture the spirit of Nuragus are those that grow on the Campidano Plain around Nuoro, Cagliari, and Oristano in the south, particularly Pala's "I Fiori" bottling.

CANNONAU

Cannonau is Garnacha (Grenache in French) and arrived on the island in the fourteenth century with the House of Aragon—though like all origin stories, this one is disputed. The most important and widely grown red grape on Sardinia, it prefers the sandy soils of the southern half of the island, in and around Nuoro,

Cagliari, and Oristano. Cannonau is a grape eager to ripen and thus care must be given lest the wines max out with 16 to 17 percent alcohol.

However, when grown at high altitude on granite and harvested appropriately, as on the best parcels, Cannonau yields a lightly colored red wine that is nevertheless deep in body, with rich minerality and herbal notes. A friendly grape, Cannonau is bottled as a monovarietal but also benefits from company, which often includes other native grapes like Bovale, Carignano, and Malvasia Nera.

CARIGNANO

Carignano probably arrived with the Spanish in the fourteenth century, though some think it actually descends from another local Sardinian variety. In Spain, the grape is known as Cariñena. Here it grows, often ungrafted, on the sandy soils in and around Sulcis on the southwestern edge of the island. What commends the grape is its ability to withstand intense heat and drought while nevertheless yielding wines with good acidity and tannins, especially when it is, as it is here, grown in sandy soil and buffeted by the sea and its winds. In fact, these harsh conditions serve to stay the grape's more fecund tendency and as a result, yields are kept low and the wines burst with dark fruit, lifted by spice. I particularly love the Cardedu, which is ripe but not overbearing, and made particularly fetching with its under-$20 price tag.

VERNACCIA DI ORISTANO

Flor, like noble rot, is a natural occurrence that blesses only the most auspicious of grapes. The naturally occurring yeast is often found in the lower valley of the Tirso River as it flows into the Bay of Oristano, where Sardinia's most intriguing grape is grown. Distinct from Vernaccia di San Gimignano, Vernaccia di Oristano is often made into nonfortified almost sherry-like solera wine. (In this, it is not dissimilar to Marco de Bartoli's Vecchio Samperi Marsala.) And upon these open vats form the thin white layer of yeast, flor.

The resultant flor-kissed wines burst with nutty almond and hazelnut flavors, sing of sea salt, and tingle with acidity. Today, sadly, plantings of Vernaccia di Oristano are declining. Nearly half of the farms that grow the grape are under one hectare, and over a quarter are more than fifty years old.

Nonetheless, though Vernaccia di Oristano is certainly a wine not in step with its time, it is timeless and, like all solera, an enchanting and fascinating mixture of history and the present. The best examples of Vernaccia di Oristano come from Silvio Carta and a small producer named Francesco Atzori.

Winemakers to Know

CARLO DEPERU AND TATIANA HOLLER OF DEPERU HOLLER

The best Vermentino I've had in Sardinia comes from this fifteen-acre vineyard, ten miles from the sea on the northern part of the island. Here Carlo Deperu and his wife, Tatiana Holler, tend to their organic vineyards, which benefit from both the mistrals of the sea and the cooling breezes of nearby Lake Coghinas, as well as remarkably variegated soil. Deperu, who is Sardinian, and Holler, from Brazil, are relatively new entrants in the Sardo wine world, having met at university in Milan and moved back to Deperu's hometown, Perfugas, in 2007. They released their first vintage only in 2008 but have proven adept in capturing the creamy, yet mineral character of the best Vermentino.

As well as this noble variety—bottled as a monovarictal in their wine Fria—the couple have taken up the cause of lesser-known varieties, including the red Muristellu and the whites Arvesiniadu and Nasco.

GIOVANNI MONTISCI

The pinnacle of Cannonau is a town in the mountainous center of the island, Mamoiada, in the province of Nuoro. Here, on two hectares of sand and granite soil at 640 meters above sea level, former mechanic turned cult winemaker Giovanni Montisci vinifies the grape from sixty-to-eighty-year-old vines, spontaneously fermented, unfiltered, and unfined. Most of the work happens in the vineyard and Montisci, a man obsessed with detail, does not shy from the labor his vines demand. They are trained *in albarello* and hand-harvested. Montisci has only been bottling Cannonau since 2004, but before that, he had been experimenting for fifteen years, selling the wine *sfuso*, or in casks.

FRANCESCO ATZORI

Francesco Atzori is one of only six producers of Vernaccia di Oristano, and one of its best. From his small farm on forty hectares of land on the western coast of Sardinia, he makes Vernaccia di Oristano and nearly as much olive oil. In both endeavors, he is minimalist by nature. He doesn't till the soil or use herbicides or pesticides. In the cantina too, he barely intervenes as the wine ages for ten years in old chestnut barrels. Each year, Atzori releases only 800 cases, which, despite the style's relatively niche appeal, quickly sell out.

Alessandro Dettori is something of the Don Quixote of Sardinian wine. Undisputedly, he is *the* pioneer of the island's natural wine scene, having begun his quest on his great-grandfather's three hectares of ungrafted Cannonau in Romangia, in northern Sardinia, in 1998. Today the acreage has expanded eleven-fold, but Dettori still insists on vinifying his flagship Cannonau "Dettori"— which comes from the original three hectares—at 16 to 17 percent alcohol. Though I've always admired his methods—hand-harvesting, biodynamic farming, spontaneous fermenting, aging in concrete, and the like—these wines are a bit too ripe for my taste. I find his other offerings, such as a lightly macerated Vermentino, more enjoyable. However, like some wines, the man himself has mellowed, a softening he chalks up to marriage and fatherhood, and his bottles too have become less overpowering.

More Exceptional Producers in Sardinia

Alberto Loi
Angelo Rivano
Capichera
Cardedu
Contini
Pala
Piero Mancini
Silvio Carta
Vigne Rada
Vigne Surrau

Sicily, Pantelleria & the Aeolian Islands

Sicilia, Pantelleria, e Isole Eolie

In 1787, when the German poet Goethe wrote, "To have seen Italy without having seen Sicily is not to have seen Italy at all, for Sicily is the clue to everything," Sicily wasn't yet part of Italy. In fact, the unified kingdom didn't exist. The island, off the southwestern coast of the toe of the Italian Peninsula, belonged to the House of Bourbon, a branch of the Spanish royal family. But the Bourbons were just the latest, and not the last, rulers who attempted to control Sicily, the largest island in the Mediterranean and one that, separated from Tunisia by the narrow Strait of Sicily, has always been of immense strategic import.

From the Greeks to the Carthaginians, Romans, Byzantines, Moors, Vikings, and Normans and through to the French and Spanish, Sicily has been fought over for millennia. As Giuseppe Tomasi di Lampedusa, one of its greatest novelists, wrote, "For over twenty-five centuries, we've been bearing the weight of superb and heterogeneous civilizations, all from outside, none made by ourselves, none that we could call our own." Nor could any of these forces truly call Sicily their own—not even modern Italy. But Goethe was right about one thing: Sicily *is* the clue to everything.

The largest of Italy's twenty regions—and a semiautonomous one, with not only its own customs and language but also its own laws—Sicily is a wild, thrilling, energetic land. The island is dominated by Mount Etna on the eastern coast, by fields of grain in the central hills, and by salt pans, cliffs, and vineyards to the west, and every vista is breathtaking. Every image is supersaturated, bristling with preternatural beauty. Even the cities—where the accumulation of kingdoms yields a joyous jumble of architectural styles, some of which are covered by graffiti scrawled like kudzu—are poetry.

In terms of wine, there is no more exciting or fertile ground. The potential for high-quality wine, long pent up, is explosive. Some of Sicily's winemakers are centered on the slopes of Mount Etna. In Vittoria, in the southeastern quadrant, another group of Sicilian pioneers, led by Giusto Occhipinti of the groundbreaking winery COS, celebrates the country's array of native grapes, while Giusto's niece, Arianna Occhipinti, a young force of nature, has become a household name with her resurrection of Frappato, the aromatic red that seems airlifted from northern Italy but is somehow distinctly Sicilian. In the far west, closer to Africa than to mainland Italy, winemakers like those at Marco de Bartoli

Frank Cornelissen (*right*)

are reclaiming Marsala: by reputation, a treacle-sweet fortified wine; by tradition, a noble one; and by innovation, a brilliant one. All across the island, winemakers unmoored by tradition and inspired by this vast and varied land are experimenting with Sicily's native grapes and making some of the world's most engaging wines.

It wasn't always this way, of course. One could—and others have—unspin the sad story of why this island was known for so long for its bulk wines, shipped by tanker to Europe for blending, and the much-maligned Marsala, destined not for drinking, but for cooking. One could delve into how, for instance, Sicily boasts the most acres under vine in all of Italy and yet the market for its consumption barely exists—and surmise perhaps that Sicilians have prized economic progress over even their own tastes. And who could blame them? Like so much in Italy, Sicily's is a tale of misguided economic incentive; like so much in Sicily, its wine problems have been added to by federal economic discrimination, a long history of failed land reform stretching back to the Roman feudal *latifondi* system, and the insidious destabilizing presence of the Cosa Nostra, a problem with which the island still struggles. But instead, the shadow of the erupting volcano, sending its plume of smoke skyward, spurs us to look not toward the past, but toward the future. And the future is thrilling.

According to Greek mythology, Mount Etna was the forge in which the Cyclops Polyphemus and the god Hephaestus made Zeus's thunderbolts. Its bulk was formed when Athena cast out the giant Typhon in the War of the Giants and its eruption an expression of his rage. Archaeologists pinpoint the first eruptions to five hundred thousand years ago on the seam between the African and Euroasian tectonic plates. It is the largest volcano in Italy, and the most active. For natives of Catania, a city in its shadow, Mount Etna is a constant presence, a source of pride and fear. They call the mountain "Mamma," which says, I suppose, a lot about both the volcano and their mothers.

Whether or not Mount Etna, the highest mountain south of the Alps and just off the coast of Africa, is a benign or malign force for its neighbors, its slopes are ideal for wine. They see large swings of temperature between day and night (diurnal shift), and the soil is a mixture of lava, volcanic ash (*tufo*, or tuff), and rock. The western slopes, buffeted by the sirocco winds fresh from the Sahara, are often too hot to be cultivated, but on the northeastern side, red grapes like Nerello Mascalese flourish, and on the east, white grapes like Carricante thrive. Both of these, in my opinion, are among the best grapes of all Italy. Whether on lava, ash, or rock, grapes here have had to dig deep to survive, and yet, this unfriendly soil has proved their salvation. As inhospitable as it is to vines, so too has it been resistant to the bane of phylloxera, and for that reason, one can find not only vines hundreds of years old here but also native grapes, many ungrafted, that were spared decimation.

In the last years of Sicilian independence, Etna boasted a flourishing wine industry. But the disembowelment of the island's economy proved nearly fatal. Only in the last twenty to thirty years has quality winemaking returned. In the 1990s, this was pioneered by Massimiliano Calabretta, Giuseppe Benanti, and Salvo Foti, Benanti's winemaker. In the early aughts, three foreigners—a Belgian named Frank Cornelissen, an American named Marco de Grazia, and a Tuscan named Andrea Franchetti—arrived, each seeing in the steep volcanic slopes potential for their own vision. Cornelissen's was of wild individualism, de Grazia's of softness and elegance, and Franchetti's of deep power. The mountain obliged. A little later, a winemaker named Alberto Graci returned from Milan to his family's native land. Now everyone wants a piece of the volcano.

Etna is almost like a vertical city in the sense that each parcel contains a story. Though the terrain is, obviously, volcanic, from plot to plot, winemakers are relentlessly pursuing their own ideal of viticulture and vinification, so it is nearly impossible to delineate an "Etna-style" wine. For that reason—and somewhat curiously for such a historically overlooked area—the slopes of Mount Etna have traditionally been divided into steeply terraced plots called *contrade,* and more recently, these *contrade* have been treated and bottled separately, as crus are in Burgundy or Barolo.

Though Etna towers over Sicily, it does not constitute the entirety of its wine renaissance. In the north, hugging the Strait of Messina, producers like Palari and Bonavita are making elegant wines from the limestone-and-clay hills near the town of Faro. At the same time that Giuseppe Benanti was experimenting with Nerello Mascalese on the vertiginous volcanic terrain of Mount Etna, three young friends were devoting themselves to organic viticulture, native grapes, and minimalist winemaking in Ragusa, a province in the southeastern corner of Sicily. On a tiny three-hectare *contrada* just outside the town of Vittoria, Giusto Occhipinti, Titta Cilia, and Rino Strano's COS began pursuing a perfect Cerasuolo di Vittoria, a red wine made from 60 percent Nero d'Avola and 40 percent Frappato. Without training, but with open minds, the friends experimented with barrique and international varieties but quickly settled on producing their wines organically, biodynamically, and in amphorae. Their wines are noble expressions of Cerasuolo di Vittoria, with so little done to obscure the wine's origins that one can almost see the red soil of Vittoria in each bottle. It was their work early on that, in conjunction with that of de Bartoli in the west and Benanti and Foti on Etna, buoyed Sicily's reputation. Their work goes on today as the wines continue to reach new heights.

But COS wasn't the only gift the Occhipintis gave Vittoria. Giusto's niece Arianna Occhipinti (alongside Elisabetta Foradori and Emidio Pepe, one of the three patron saints whose portraits greet diners at Fausto) embodies all that vino vero can be. At her winery, founded in 2006, Arianna has focused her intense creativity and dedication toward Frappato, a red grape that does well on the *terra rosso* (red earth) of the southeast, as well as Nero d'Avola, the grape hailing as it does from the Ragusan town Avola, and a white wine made with Albanello and Zibibbo. From Arianna's vineyards comes a beautiful monovarietal wine called simply "Il Frappato," which is a near-perfect portrait of a grape; a deep yet soft Cerasuolo di Vittoria; and exemplars of Nero d'Avola.

The third vanguard of Sicilian wine lies on the northwestern coast, in and around Marsala and Trapani. In the 1970s, a former race car driver with a serious chip on his shoulder named Marco de Bartoli was intent on rescuing Marsala from its ignominy. The wine, traditionally made with a mixture of Ansonica (called here Inzolia and the same grape found in Maremma) and Grillo, was by law fortified with additional spirits and then with cooked grape must, for color and sweetness. That wasn't the invention of Marsalans, but of an English trader named John Woodhouse, who "discovered" the similarity between the native wines of Marsala and the sherry so beloved in his home country. Both are made with the solera method, wherein various vintages are aged together and constantly replenished. And both can be vinified as dry or sweet wines.

For a few centuries, Marsala had been held in high regard as a sweet wine or digestif. But by the time de Bartoli was born, it had become a cheap industrially made product so adulterated by additives that any connection to its namesake village was gone. One tasted alcohol and sugar and little else. That had never sat well with de Bartoli, who used to help his grandparents in their vineyard when he was young. Though his family had grown to be one of the largest producers of industrial Marsala, de Bartoli had always suspected the wine was capable of more than what his family was making. Yet his family was as resistant to change as he was to compromise. They fought, and he fled. But after leaving Sicily to pursue a career as a race car driver, de Bartoli returned, taking over his family's *contrada* in Samperi, just outside Marsala. He began soliciting solera barrels from old winemakers—with which they were too happy to part—and experimenting with a high-quality Marsala. Well, it was high quality, but it wasn't Marsala, because the regulations demanded the addition of other alcohol. Like any just man when confronted with legislative idiocy, de Bartoli didn't budge. His 100 percent Grillo wines, including his flagship "Vecchio Samperi," grown organically and aged in oak and chestnut barrels, can't be labeled Marsala, but they are perhaps the truest expression of his beloved terroir. His angry rejection of the status quo helped Marsala shed its debased image as a wine fit only for cooking, and today it, at least in the hands of de Bartoli, shows itself a wine worth not only being drunk but savored. Thanks to the work of de Bartoli, who died in 2012 at only sixty-six years old, an increasing number of wineries in the far west of Sicily have taken up the banner of Marsalan restoration; they are few but growing. And after them, the deluge, one hopes, of inferior wine will subside, leaving only the true believers.

To the far west of Sicily, the island Pantelleria, which sits between Africa and Sicily, is truly the wild, wild west of Italian wine. (Though, to call this arid volcanic land forty-five miles from Tunisia Italy feels somewhat incongruent.) Along these hills, exposed to endless winds and brilliant sun, vines are trained to grow *in albarello*—as little bushes—and are dug deep into the volcanic soil to protect themselves from the wind. Traditionally the island was known for its sweet wine, Passito di Pantelleria, made with air-dried Zibibbo (the name means raisin in Arabic); the grape is also known as Moscato d'Alessandria, the mother of Grillo. But it produces some dry white wines as well, called Moscato di Pantelleria, with undried Zibibbo, which through the constant buffeting of the wind, retains its acidity. It was with this wine, according to Greek mythology, the goddess Tanit seduced Apollo.

Among Zibibbo's greatest advocates is the island's most important producer, Donnafugata, whose dessert wine is called "Ben Ryé," Arabic for "son of the wind." But we have Marco de Bartoli, who purchased land here in 1980 and kick-started the reexamination of Passito, to thank for its current renaissance. He is joined by the natural wine–world darling Gabrio Bini and an exciting new producer, Abbazia San Giorgio, who are experimenting with skin maceration, among others in Pantelleria.

Unlike the rest of Sicily, on the island of Pantelleria, winemaking has been on a downward trajectory. In the 1930s, there were nearly 15,000 acres of vines planted on the island; today there are only 3,700. But I'm confident that as these wines gain traction, winemakers will return to the steep, windy slopes, to the hard work of growing there, and to the mission of expressing the beauty of their land.

More vibrant, for now, are the Aeolian Islands, northeast of Etna—Lipari, Vulcano, Salina, Stromboli, Filicudi, Alicudi, and Panarea. Each is blessed with volcanic soil, and on each island, winemakers are awakening to the possibilities of high-quality wines from their vines. The islands are named after Aeolus, the god of wind, and it is Aeolus who makes possible their most important wine, a Passito called Malvasia delle Lipari.

These wines, made by drying Malvasia grapes on straw mats in the sun to allow the wind and heat to concentrate the sugars, are abundantly sweet but with a strong backbone of acidity. Though Lipari, the most inhabited island, gives the wine its name, most production comes from the island of Salina, known for its caper trade. The first Malvasia delle Lipari to emerge from the Aeolian Islands was that of Carlo Hauner, in the 1980s, who aged the wine in barrique and exported it internationally. The boom triggered by Hauner has since cooled to a murmur, but his son, also Carlo, and a host of other producers are experimenting with dry styles of Malvasia, which are sold as IGT Salina wines. Though these islands lack the altitude of Etna, they do not lack potential. Now hotbeds of experimentation, they produce nearly 300,000 bottles a year, some of it very good.

Native Grapes

CARRICANTE

If the King of Etna is Nerello, the Queen is Carricante, one of Italy's greatest white grapes. Unlike Nerello, which grows mostly on Etna but is found elsewhere, the domain for Carricante is a scant 146 acres on the southern and eastern slopes of Mount Etna. The grape reaches full ripeness while maintaining low levels of alcohol and high acidity, making it a prime candidate for aging. But it demands work. Though productive—the name means loading up, for what was done to donkeys with the loads of fruit the grape is capable of producing when unrestrained—Carricante demands high-altitude plantings, preferably at between 950 and 1,050 meters above sea level, at heights so vertiginous that not even Nerello can bear.

Benanti's Etna Bianco Superiore "Pietra Marina" is the archetypical expression of this grape; steely and taut in its youth, rounder with age. But the wine's capability of expression is still being explored. Alberto Graci's Etna Bianco "Quota 600" is intriguingly textured, and Girolamo Russo's Nerina Etna Bianco is a flinty, mineral white wine with a joyful thread of fruit. Though still in its infancy in terms of vino vero vinification, I believe Carricante may be the Italian white wine to give Burgundy a run for its money and at a fraction of the price.

GRILLO

The western coast of Sicily has lagged behind the east, having long been given over to the production of bulk Marsala wine, a blend made of Aglianico—here called Inzolia—and Grillo. Only in the last twenty years have winemakers, following the trail blazed by Marco De Bartoli, devoted themselves to reclaiming the blend.

Overlooked during the years of sweet industrial oblivion was Grillo, a white grape of enormous potential. If not squandered in its ensemble casting in Marsala wines, Grillo shows itself capable of wines of great depth and minerality. The best Grillo wines are still produced at Marco De Bartoli. These include the fresh and lively Vignaverde; the more complex and textured Grappoli del Grillo; and the unique and multifaceted "Terzavia," a sparkling iteration of Grillo made using the *metodo classico*. De Bartoli had been joined by a pack of other stellar winemakers in Marsala, including Nino Barraco, who produces Grillos both bright, salty, and acidic (Vignaverde) and rounder and textured (Terre Siciliane

Grillo), as well as by non-Marsalan winemakers like Aldo Viola, whose Grillo enjoys skin maceration; Elios, which makes both white and orange versions; and Viteadovest, which blends Grillo with Catarratto for a fascinating wine.

CATARRATTO

The name Catarratto means cataract, as in waterfall, and refers to the prodigal amounts of wine made from what is the most common grape in Sicily and the second-most common in all of Italy. (The first is Trebbiano Toscano.) The propensity of the white grape to yield immense amounts of wine is responsible for both its omnipresence and the low regard in which it has been traditionally held. Grown all over Sicily with the exemption of Etna, Catarratto was long used in the production of Marsala as a cheaper alternative to Grillo. (In fact, Catarratto—along with Moscato d'Alessandria—is the parent of Grillo.)

But those same saviors of Marsala and friends of Grillo have taken up Catarratto's cause. Today producers like De Bartoli and Barraco are bottling single-varietal expressions. In the case of the former, the result is high-toned and herbaceous, with a creaminess earned from seven months of aging on lees; in the latter, a wine full of herbal and seaweed flavors that transport you instantly to a seaside café. Some of the best also include the textured white expressions of Francesco Guccione and the orange wines of Bosco Falconeria.

NERELLO MASCALESE

The king of Etna's red grape scene, Nerello Mascalese, a grape of intense minerality, grows on the north, south, and eastern slopes of the volcano. To my mind, it is one of Italy's top grapes, as capable of expression as Nebbiolo but with the added benefit that so many very old and ungrafted vines are still extant. Wines derived from this thin-skinned variety, often paired with Nerello Cappuccio, as close to a soulmate a grape can have, are redolent of sour cherries, tobacco, smoke, and wild herbs. That Nerello Mascalese is so often compared to Burgundy reveals more about our innate need to frame the unknown within the boundaries of the known than any inherent qualities. It is sui generis, nothing but itself, and it needn't be more.

Fortunately for fans of this grape, the wildly varying modalities of Etna's winemakers offer it varied expression: from the elegance of Romeo del Castello Etna Rosso "Vigo" to the power of Salvo Foti's Etna Rosso "Vinupetra," made from century-old vines, to the soft refulgence of Girolamo Russo's luxuriant expression, Nerello Mascalese shines.

NERO D'AVOLA

Though Nerello Mascalese reigns over the volcano, Nero d'Avola rules much of the rest of the island. The grape was once sent abroad in industrial tankers to pad out anemic reds from the rest of Europe. Now, happily, it makes that trip finely expressed in estate bottlings with the respect it deserves.

The second-most widely planted variety in Sicily, behind the white Catarratto, Nero d'Avola expresses itself so darkly that the wine made from it is sometimes called simply *vino nero*. With swaggering ripeness, intense fruit, and the affordability of the undiscovered, Nero was *the* wine of the moment in the late nineties and early aughts. But as with pop stars and gold mines, a host of opportunistic producers flooded the zone, producing exorbitant quantities of the grape, muddying the waters, and somewhat dulling its clout. Now it's on the rebound again. Winemakers like Arianna Occhipinti in Vittoria have drawn from the late-ripening grape's high-toned tender, floral notes, while in the nearby Noto Valley, the grape's saline notes sound in the wines of Marabino. In the capable hands at Gulfi, the grape yields four single-*contrada* bottlings of finely textured Nero d'Avola, each wonderfully distinct.

FRAPPATO

Traditionally Frappato, a native red grape that hails from the southeastern corner of Sicily near Ragusa, has been blended with Nero d'Avola in Cerasuolo di Vittoria, Sicily's only DOCG wine. Though the blend benefits Nero d'Avola, to whose depth and body Frappato adds vivacity and aroma, all too often Frappato is subsumed. When left on its own, the thin-skinned aromatic grape yields a light-colored, moderate-alcohol, and medium-bodied but nevertheless refreshing red wine. It's hard not to admire the pluck of Frappato. That it hails from a region south, latitudinally, of Tunisia and not, as one might suspect from its almost Alpine spirit, from the north, adds to its already considerable charm.

Among the best examples of Frappato are those made by various members of the Occhipinti family. At COS, Giusto Occhipinti's vineyards, the wine bears cherry-spice and floral notes. At Arianna Occhipinti's namesake winery, those floral aromas are more restrained and herbs and plums are in the foreground.

The unblended Frappato scene is rapidly expanding, with wineries like Manenti, Lamoresca, and Portelli offering outstanding wines.

Winemakers to Know

SALVO FOTI OF I VIGNERI

Like Giulio Gambelli in Tuscany, Salvo Foti is both an influential consultant and an important winemaker in his own right, and one who has helped Etna find its own identity. Foti, a native of Etna, has long worked with and advocated for native grapes like the Nerello brothers—Mascalese and Cappuccio—and quality Catarratto. He is a master of extracting from the difficult volcanic soil and an advocate for training vines *in albarello*, and his influence has been felt in the many vineyards across the island for whom he's consulted, among the winemakers he inspired, and, of course, at I Vigneri, his biodynamic farm and vineyard just outside Catania, established in 2001.

FRANK CORNELISSEN

The son of a Belgian wine broker, Frank Cornelissen scaled the slopes of Mount Etna in the early 2000s and has never come down. Since 2001, when he founded his vineyard on less than half a hectare on the northern slope of Etna, he has been relentless in his devotion to the volcano and its vines. Though always farming organically and hand-harvesting—there is actually no other option on these steep slopes—Cornelissen refuses to be tethered by expectations. (In this way, he reminds me of Josko Gravner: The wine is all that matters.) For Cornelissen, this means constant experimentation and evolution. He uses not just amphorae but also massive epoxy containers. It means, in an act of hetero-heterodoxy, returning to filtering his wines. It means that wines he made ten years ago look nothing like the wines he makes today. (Today's wines, happily, tend to be much more consistent.) But no one can argue with the tremendous results. His signature estate bottlings, Susucaru and Munjebel Bianco and Rosso, are idiosyncratic. So widely admired and, as a result, frequently counterfeited are his wines that in 2016 he began using RFID microchip technology to ensure the authenticity of his bottles.

ANTONIO AND SALVINO BENANTI OF CANTINA BENANTI

When in 1988, Giuseppe Benanti Jr. decided to revitalize his family's nineteenth-century vineyards on the eastern slopes of Etna, no one knew of what heights those slopes were capable. That included Benanti himself, a chemist and an

Frank Cornelissen

executive at the pharmaceutical company founded by his father. So he hired a slew of professionals hailing from everywhere from Burgundy to Asti, as well as the young Etna specialist Salvo Foti. The group performed endless micro-vinifications seeking out the ideal grapes for the land. (Benanti was well versed in clinical trials.) The first vintages, a red called Rovittello and the Etna Biacno Superiore "Pietra Marina," were released in 1990 and quickly established Benanti as one of Etna's heroic winemakers. He had tamed the volcano. Even today, "Pietra Marina" is still widely considered one of Italy's greatest white wines.

Within a few years, Benanti had acquired vineyards on every side of the mountain. They continue to express Etna's variegated terroir with single expressions of the long-overlooked Nerello Cappuccio, as well as Carricante and Catarratto. Today the company is run by Giuseppe's twin sons, Antonio and Salvino, who have grubbed up international grapes, off-loaded lower-quality vineyards, and focused the winery on even higher-quality wine. Foti, meanwhile, has been replaced by another talented winemaker named Vincenzo Calì. Always deeply invested in pushing Etna to more rarefied heights, the Benantis completed a rigorous study of native yeasts, which they now use exclusively in their cellars, and continue to rid themselves of French grapes.

GIUSEPPE RUSSO OF GIROLAMO RUSSO

The Russo family has grown Nerello Mascalese on the north side of Etna, in the town of Passopisciaro, for as long as Giuseppe Russo can remember. But his father, Girolamo, a charcoal seller, sold his grapes in bulk. After Girolamo died, the younger Russo, a pianist and music teacher, decided to plumb the deeper tones of these vines. Inspired by his fellow Etna pioneers, Russo vinifies his Nerello Mascalese by individual *contrada*—in descending altitude, San Lorenzo, Feudo, and Feudo di Mezzo—as well as making a wonderfully vervy white with 70 percent Carricante called "Nerina," after his mother, and a 100 percent Nerello Mascalese rosato.

CHIARA VIGO OF ROMEO DEL CASTELLO

Etna's lava flow both sustains and threatens the vines of Chiara Vigo and her mother, Rosanna Romeo, the extraordinary winemakers behind Romeo del Castello. Though vines had long grown on the farm, on the north side of Etna and inherited by Rosanna's mother, its grapes had been sold off for industrial use. In 1981, a massive eruption from Etna sent great flows of lava toward the farmhouse and while the house itself was saved—by either divine providence or luck—most of the vineyards were not except, almost miraculously, a hundred-year-old patch of Nerello Mascalese. Even more fortuitously, as the lava hardened, it formed a wall that now protects these vineyards like a natural terrace. For twenty years, Rosanna tended the farm alone after her husband died and while her daughter traveled the world, earned her master's, and became a yoga teacher and an author. But in 2007, after a chance encounter with Salvo Foti, Vigo returned to the vineyards and, in turn, returned vigor to the vineyards. Today it is Vigo who makes a minute amount of Etna Rosso from the Randazzo vineyard 700 meters up on the northeast slopes of Etna. Her Etna Rosso is called Allegracore, which means, literally, the place that makes your heart happy. Never has a cuvée been more aptly named.

ANNA MARTENS AND ERIC NARIOO OF VINO DI ANNA

After traveling around the world from Chile to South Africa, the itinerant Australian winemaker Anna Martens was looking for adventure when, in Tuscany, she met Eric Narioo, a natural wines importer based in London. Martens, who had spent four years making the Super Tuscan Ornellaia, took up Narioo's offer to embark upon a journey through the Jura, Loire, and Savoie. The pair fell in love and, both being oenophiles, decided to make wine together naturally.

They moved to Etna in 2007 (though Eric still has his import business in London), when there were still many abandoned vineyards on the slopes. In 2010, they purchased a six-hectare parcel of land on the north slopes of Etna, full of old-growth Nerello Mascalese, Catarratto, and Grecanico. Since then, they've increased their holdings to twelve acres (and two children) and are producing a range of earthy pure-spirited wines. These, all organically grown and naturally fermented, include some aged in Georgian amphorae or utilizing nearly unknown varieties, like Bianco G, made with 100 percent Grecanico Dorato and bursting with minerality and wildflower blossoms.

EDUARDO TORRES ACOSTA

One of the newest winemakers in Etna and already one of the most acclaimed, Eduardo Torres Acosta grew up on the volcanic Canary Islands, interned with Arianna Occhipinti in 2012, and, in 2014, began vinifying his own bottles. For those vintages, he relied on other people's fruit—by that time, land was scarce on Etna—and on Occhipinti's cellars. But finally, in 2014, he found a parcel from a single *contrada* high up on the north side, on which he grows Nerello Mascalese as well as Minella, a rare native grape. His wines, though vinified with minimal intervention, are nevertheless precise expressions of Etna Bianco and Etna Rosso.

Arianna Occhipinti

GIOVANNI SCARFONE OF BONAVITA

In 2005, when Giovanni Scarfone returned home to Faro, a little-known region in Sicily's extreme northeast corner, he came equipped with a degree in agricultural science from the University of Bologna and the desire to turn his family's small five-acre vineyard into a business. But Scarfone quickly realized that the best strategy was to abandon much of the technical wizardry he'd learned in the classroom for a more bare-bones, back-to-nature approach. Though both Nerello Mascalese and Nerello Cappuccio flourish in Faro, so too does a third little-known variety called Nocera, a red with bright-edged acidity. These grapes, protected and cooled by the forests ringing the area, grow in white clay soil, in a much more moderate climate than on Etna's slopes. This leads to longer maturation and terrific depth. Traditionally, these three grapes were vinified by *contadini* into a light, lively rosé meant to accompany the full arc of dinner. It is this tradition Scarfone follows with his Bonavita rosato, a deep-colored rosé made from eighty-year-old vines and in almost-panic-inducing scarcity.

GIUSTO OCCHIPINTI OF COS

Since founding COS with two friends in 1980, Giusto Occhipinti has never stopped evolving his practice. After a period of experimentation—from oaks in various sizes and degrees of neutrality to varieties with different levels of indigenousness—COS, now primarily run by Occhipinti, has settled into a rhythm of producing exemplary Cerasuolo di Vittoria in large Spanish amphorae, textured whites, and outstanding Frappato. The amphora wines, dubbed Pithos, are among my favorites in Italy and clear quantum leaps for Sicilian wine. COS also puts forth excellent blends of native white grapes, including one made with Inzolia and Grecanico Dorato called Ramì, a wine with rare depth and minerality even at a lowly 11.5 percent alcohol.

ARIANNA OCCHIPINTI

Arianna Occhipinti embodies the future of Sicilian winemaking in particular and, one hopes, Italian winemaking in general. Occhipinti's trajectory has been steep. She started working with her uncle, Giusto Occhipinti, at age sixteen, studied oenology at the University of Milan, returned to Vittoria and began experimenting on an acre of abandoned vines near her family's house, and then launched her own winery in 2006. Over the past fifteen years, Occhipinti has almost single-handedly brought Frappato back to life, much in the same way her uncle championed Cerasuolo a few years earlier. Not only does she make some of the cleanest, most expressive versions of the grape—vinified in three single-*contrada* variants—but her own Cerasuolo is among the best. While she's become a rock star in the world of natural wine, Occhipinti prefers to think of herself as a lover of her land. Though her grapes—mostly Frappato and Nero d'Avola—are hand-harvested, are naturally fermented, and never see oak, all that is simply a means for a more crystalline expression of her beloved Vittoria.

THE DE BARTOLI FAMILY OF MARCO DE BARTOLI

Simply put, Marco de Bartoli was the savior of Marsala. And his influence extended even farther west, to the island of Pantelleria, where, in the 1980s, he was one of the first to pioneer the resurgence of Passito di Pantelleria.

Among his many innovations was his almost truculent resistance to parting with his beloved Marsala. Even today the old cellar underneath the de Bartoli house is filled with what appear to be infinite barrels, jealously guarded. But all treasure is hard to wave goodbye to, and treasure this is. The barrels contain the life's work of de Bartoli, who pioneered an old style of Marsala called Vecchio, a nonfortified *vino perpetuo*. Though known mostly for Marsala, the restless de Bartoli also experimented with sparkling wines and orange wines. Today his three children push the de Bartoli legacy forward with a range of precisely made wines from indigenous grapes in a variety of styles, including *metodo classico*, white, orange, red, dessert, fortified, and more.

NINO BARRACO

Trapani, just north of Marsala on Sicily's western edge, is famous for its salt pans that lie along the Mediterranean Sea. But just beyond them are the vineyards of Nino Barraco, whose grapes hold their marine character tight all the way through to the bottle. Since he started the winery in 2004, Barraco has specialized in two of Sicily's greatest but underrespected white grapes: Grillo and Catarratto. His "Vignammare," a 100 percent Grillo wine from grapes grown just meters from the sea, bristles with lemon zest, sea salt, and mineral waters. His Catarratto draws smokiness from its volcanic soil and a fine ocean mist from the sea before it. Barraco, who says his goal is "not to create a perfect wine but one that pours with emotion," has expanded his holdings to ten hectares, from which he makes an array of stellar wines, from a dry Zibibbo to a zippy rosato from Nero d'Avola, called Rosammare, to a rare monovarietal Pignatello. But it is his "Vignammare" that remains to me the perfect encapsulation of this elemental land.

Nino Barraco

Sebastiano and Gipi De Bartoli

More Exceptional Producers
in Sicily, Pantelleria,
and the Aeolian Islands

AEOLIAN ISLANDS

Azienda Agricola Caravaglio
Carlo Hauner
Tenuta di Castellaro

ALCAMO AREA

Azienda Agricola Elios
Il Censo

ETNA

Biondi
Calabretta
Etnella
Graci
I Custodi delle Vigne dell'Etna
I Vigneri
Monterosso
Palmento Costanzo
Scirto
Tenuta delle Terre Nere

FARO

Palari

MARSALA AREA

Aldo Viola
Francesco Guccione
Viteadovest

NEAR VITTORIA
(NOTO AREA)

Gulfi (their Carricante, not
grown on Etna, is great!)
Marabino

PANTELLERIA

Abbazia San Giorgio
Marco De Bartoli
Serragghia (Gabrio Bini)

VITTORIA

Lamoresca
Manenti
Portelli

Further Reading

Bookshelves groan under the weight of all the words spilled on wine. Some of these tomes are indeed worth delving into. Anything by Ian D'Agata, a savant of Italian wine whose grasp of Italy's native grapes, as evidenced in *Native Wine Grapes of Italy* (2014), is truly encyclopedic. As mentioned earlier, David Lynch and Joe Bastianich's *Vino Italiano: The Regional Wines of Italy* is a must-read, though in need of an update after fifteen or so years. But to be honest, what has been most helpful in the writing of this book is the tremendous amount of research and work that has been done by many of these wines' importers. I've spent more time than I wish to calculate on the sites of Louis/Dressner (louisdressner.com); Polaner Selections (polanerselections.com); and Zev Rovine Selections (zrswines.com), and many more hours deep-diving into the producers they carry. Frequently the profiles they offer include virtual site visits and always a breakdown of each wine. Sometimes, as is the case with Jules Dressner's reports, the writing is terrific, funny, and truly informative.

Another resource, which didn't exist when I was just starting out, are the websites of the wineries themselves. Often available in English (and Google Translate does a bang-up job if not), these offer in-depth histories and often fanciful, poetic evocations of the winemaker's philosophy. Though Italian wineries can be the worst at SEO, it's worth scrolling until you find the websites.

Finally, though it isn't reading, obviously, I've certainly found it to be true that if you are in a wine store or restaurant offering exceptional wine, whoever is responsible for the wine there is usually more than happy to talk about it. These are carefully chosen wines that can't help but transmit their passion. And we, true believers in vino vero, can't help but spread the word.

Acknowledgments

For me, wine has been as much a pursuit of knowledge as it has been one of pleasure. I'm grateful to all of the wine teachers over the years who have guided me on this journey. I'm especially grateful to Linda Lawry, who is now retired but taught my first-ever wine class at NYU and then the advanced class and diploma class of the WSET (Wine & Spirit Education Trust) at the International Wine Center. It was she who inspired my desire to learn. Thank you also to Mary Ewing Mulligan, MW, who offered me a job at the IWC and allowed me to be a teaching assistant for the diploma class, the only way I could afford it as a college student. Thanks to Ian D'Agata, who first taught me the importance of native Italian grapes. And to David Lynch who, with Joe Bastianich, wrote *Vino Italiano*—which is the book I've read more times than any other, on any subject. He was also my boss when I was twenty-three years old and worked as a sommelier at Babbo for all of three months—I learned a lot from him in that short time and continue to follow his career.

My professional career has been entirely focused on Italian wine, first in a retail Italian wine store, then with an Italian wine distributor, and finally at an Italian restaurant, before opening my own restaurants. At Italian Wine Merchants, Sergio Esposito took a chance on an underage college student and hired him to work at his beautiful store. August Cardona was my boss at IWM and gave me my first chance to own a restaurant when we later became partners in what became a slew of restaurants in downtown Manhattan. I also want to thank Dominic Nocerino and Vince Attard of Vinifera Imports. I was twenty-two when Dominic hired me to be a salesperson. Vince took me under his wing and was always incredibly supportive; he showed me that a rising tide lifts all ships. My current restaurant partners, David Foss of LaLou and Erin Shambura of Fausto, have been patient and supportive throughout the process of writing this book. Thanks also to anyone who has ever dined in any of the six restaurants in which I've had any ownership.

I could not have written this book if it weren't for the countless independent artisan producers in Italy who do the hard work every day. Thanks especially to Stefano Papetti Ceroni, Chiara De Iulis Pepe and the entire Emidio Pepe family, Elisabetta Foradori, Giovanni Manetti of Fontodi, Arianna Occhipinti, and Giovanna Morganti, who made the 2001 Podere Le Boncie Chianti Classico that first got me so excited about real Italian wine. Also to Giuseppe Vajra of the extraordinary G.D. Vajra Winery and Fabio Alessandria of G.B. Burlotto—their generosity knows no bounds. Thanks also to Mitja Sirk, who is a producer but also runs his family's fantastic restaurant and inn, La Subida in Friuli. To Josko Gravner and the late Gianfranco Soldera, for stubbornly, resolutely making the wines they believe(d) in, like so many others in this book. And though I never got to meet him, Bartolo Mascarello has always been an inspiration. There are countless other producers in Italy who have opened their vineyards and homes to me, and I thank them all so much.

I'd also like to thank the wine importers and distributors who bring so many of these exceptional wines to the States. There are too many to list, but I'm especially grateful to Neal Rosenthal and Blake Johnson; Jules Dressner of Louis/Dressner; Lyle Railsback, formerly of Kermit Lynch Wine Merchant; Ben Augustine and Jane Berg of Kermit Lynch; Jenny Lefcourt of Jenny & François Selections; Jan D'Amore of Jan D'Amore Wines; David Bowler, David Gordon, Juliette Pope, and Kevin Russell of David Bowler Wine; the team at PortoVino; Iacopo Di Teodoro of Artisanal Cellars; Robert Bohr of Grand Cru Selections; Enrique Ibañez of IPO Wines; the team at Skurnik Wines; T. Edward Wines; SoilAir Selection; Marco Irato of Tradizione Imports; Doug Polaner and Hannah Norwick of Polaner Selections; Vinifera Imports; Zev Rovine Selections; Vignaioli Selection; Panebianco Wines; Domenico Valentino; MFW Wine Co.; VOS Selections; Schatzi Wines; DNS Wines; Satyr Picks; Mucci Imports; Coeur Wine Company; and Critical Mass Selections. I've discovered many more great Italian wines by tasting them with their distributors in New York City than I ever could have in Italy. Their work, finding these wines and telling their stories to buyers, is largely overlooked by consumers but is essential.

Thanks to Antonio Galloni, who keeps inviting me back to La Festa del Barolo, where I get to taste some of Italy's greatest wines; I've been reading his writing since my days at Italian Wine Merchants. And to Ray Isle of *Food & Wine*, who always reminds me to make sure wine is fun and cheerful. Thanks also to Eric Asimov of the *New York Times*, for being so approachable, and for inviting me to the Aglianico and young Barbaresco tasting panels—he knows I can handle the tannins!

To Raj Parr, who was incredibly generous with his time. And to Jeff Porter, Charles Antin, Greg Ruben, Josh Nadel, Matt Conway, Thomas Pastuszak, Tommy Wenzlau, Aaron Sherman, Raj Vaidya, and Alex Zink. Thanks too to Bobby Stuckey, because if you love Italian wine, if you work in the restaurant industry, if you're a runner, you look up to Bobby—and I'm all three. He also took me on my first trip to Friuli, which was eye-opening, to say the least.

To my editor, Jennifer Sit at Clarkson Potter—thank you for jumping into this project headfirst, with fresh eyes. Thank you to the talented photographer Oddur Thorisson, for taking the beautiful photographs, and to his wife, Mimi, for her generous hospitality in having me to their beautiful apartment in Torino and for lending me her husband so he could run around Italy and shoot this book. And to Rica Allannic of the David Black Agency: I can't imagine there is a better agent in the world. This book would not have happened without you, and I'm so glad to have you in my corner.

To my coauthor, Joshua David Stein: JDS—I loved working on this project with you. I only wish it had taken longer—and that it didn't have to be done during a pandemic, so that we could have enjoyed many meals and wines in Italy together. Along with the winemakers, you are the hero of this book; without you, it would never have come to fruition. And, thanks to your illustrious career as a children's book author, I'm proud to say that I'm the owner of more books by Joshua David Stein than by any other author in my library.

Most of all, to Ilyssa Satter, my life partner and mother to our beautiful son, Cole. You inspire me to be better at everything I do. This book wouldn't have been half as good without you.

I want to thank Joe Campanale, who gave me a fridge full of the best wines in the world, a mind awake to the possibilities of vino vero, and the opportunity to help write a book capturing the passions of a country full of passionate winemakers. I'd also like to thank Jenn Sit at Clarkson Potter for her thoughtful editing.

Index

Library of Congress Cataloging-in-Publication Data
Names: Campanale, Joe, author. | Stein, Joshua David, author.
Title: Vino: the essential guide to real Italian wine /
Joe Campanale with Joshua David Stein;
photography by Oddur Thorisson.
Description: New York City: Clarkson Potter, 2022.
| Includes index. |
Identifiers: LCCN 2021015689 (print) | LCCN 2021015690
(ebook) | ISBN 9780593136140 (hardcover)
| ISBN 9780593136157 (ebook)
Subjects: LCSH: Wine and winemaking—Italy.
Classification: LCC TP559.I8 C273 2022 (print)
| LCC TP559.I8 (ebook) | DDC 663/.200945—dc23
LC record available at https://lccn.loc.gov/2021015689
LC ebook record available at https://lccn.loc.gov/2021015690

ISBN 978-0-593-13614-0
Ebook ISBN 978-0-593-13615-7

Printed in China

Photographer: Oddur Thorisson
Editor: Jennifer Sit
Editorial assistant: Bianca Cruz
Designer: Mia Johnson
Cartographer: Irene Laschi
Production Editor: Christine Tanigawa
Production Manager: Kelli Tokos
Compositor: Merri Ann Morrell and Hannah Hunt
Copy Editor: Judith Sutton
Marketer: Samantha Simon
Publicist: David Hawk

10 9 8 7 6 5 4 3 2 1

First Edition